Crater Lake

Gem of the Cascades

Cover Photograph: *A panoramic view of the Crater Lake caldera from along the Garfield Peak trail. Many prominent geological features on the northern caldera wall can be seen in this photograph. The background image is an early interpretation of Mount Mazama during the climactic eruption, from a painting by Rockwood in the 1930s.*

Frontispiece: *Pumice Castle rests on the caldera wall with Redcloud Cliff in the background. Portions appear to be welded, resulting in erosion into spire-like patterns that mimic the pinnacles.*

Crater Lake
Gem of the Cascades

The Geological Story of
Crater Lake National Park, Oregon
Third Edition

K R. Cranson

KRC Press
Lansing, Michigan

Dedicated to the early explorers and settlers of the High Cascades, especially those whose daily chores failed to smother their appreciation and enjoyment of places like Crater Lake. Because of their interest in and efforts to preserve unique natural places, these areas are now available for anyone to visit. And, to my great pleasure, I am able to study geology in its original and undiminished condition.

Printed in the United States of America by KRC Press, 226 Iris Avenue, Lansing, Michigan 48917.

All photographs, graphics, typesetting, and layout by the author unless otherwise noted.

Library of Congress Control Number: 2005905182
ISBN: 0-9770880-0-6 (soft cover)
ISBN: 0-9770880-1-4 (hard cover)

First Edition (May 1980) – 2,000 copies
Second Edition (May 1982) – 7,000 copies
Third Edition (May 2005) – 10,000 copies
(9,500 softback, 500 limited Commemorative Centennial Edition)

Contents

Boxes

Tables

Plates

Preface
Third Edition

Crater Lake certainly is a gem, one of the crown jewels of the United States that we call national parks. Indeed, no other country has such a heritage of national areas. Crater Lake occupies a special place in the memories of anyone who has ever visited and learned its geological story. It also holds a special place in classic geological thought and theory. Any geologist can tell you why. To a large degree, Crater Lake's fame results from its geological past and our ability to interpret that record today. That's what this book is all about.

My goal in writing and publishing this small volume on the geology of Crater Lake National Park stems from a desire to explain this wonderful bit of landscape to park visitors and others interested in the park. Although it is meant for the layperson, trained geologists may find some of the ideas refreshing. In planning, writing, and laying out the book I have attempted to keep both audiences in mind. This resulted in the abundance of photographs, diagrams, maps, tables, and other features to help illustrate Crater Lake's geology as clearly as possible. These techniques, of course, are familiar to geologists but should also be helpful to all readers.

When 7,000 copies of the second edition were printed in 1982, I thought they would last forever. After all, how many people are willing to purchase a book on the geology of Crater Lake National Park? As it turned out, I was wrong, and the book was out- of-print and out-of-date by the early 1990s.

In the more than twenty years since that edition was published, there has been an amazing amount of volcanology research conducted worldwide. This has greatly increased our ability to interpret the geology of places like Crater Lake National Park. Researchers in the park have been busy, resulting in new discoveries and a better understanding of previous work. All of this, of course, had to be digested and incorporated into this edition. Thus, preparing a third version of ***Crater Lake – Gem of the Cascades*** rapidly escalated into a major project.

Fortunately, with the great advances in self-publishing methods, a completely new format was possible. The obvious difference in this edition is the larger format, which permits more flexibility in layout. Special boxes have been created to better explain certain technical and related aspects that are not part of the actual storyline. A new appendix has also been added to assist with those geological terms that might not be part of the reader's normal vocabulary. And, finally, the reference appendix has been updated to reflect the pertinent research conducted over the past twenty-plus years.

A number of new people have assisted with this work as well as most of those noted in the second edition. As before, special thanks goes to Dr. Charles Bacon with the United State Geological Survey (USGS) who reviewed more than half the text (Chapters 4, 5, 6, 7 and 9). Thanks also to Steve Mark, National Park Service (NPS) Historian at Crater Lake (Chapters 1 and 8) and for answers to many questions; Dr. Thomas Vogel, volcanologist at Michigan State University, and Art Weinle, Master Earth Science Teacher (Chapter 2); Dr. Douglas Larson, longtime limnology researcher on Crater Lake (Chapter 8).

My wife, Sharon, deserves much of the credit for making this book happen. We've lost count of the number of times she read and corrected all the chapters and provided suggestions on how to improve them. She also provided encouragement and kept my nose to the grindstone when I wondered if the book would ever be finished. While all these people pointed out mistakes and freely made suggestions, any errors must be credited to me.

K R. Cranson
Lansing, Michigan
May 1, 2005

Chapter 1
Discovery

Late in the afternoon on June 12, 1853, a small band of gold prospectors rode slowly up a mountain in south central Oregon. They were in search of the rumored Lost Cabin Gold Mine and the instant fortune that it would surely yield. Although details are sketchy, it appears that three members of this party, John Hillman, Issac Skeeters, and Henry Klippel, were in front of the others and unaware of what existed just ahead beyond the ridge. Little did they realize what the next few hours had in store for them, and ultimately, the entire United States and the world.

These men had arrived in Jacksonville, Oregon, the previous fall. They had come for the new gold "diggings" that had been discovered in local streams. The winter of 1852–53 was especially severe, with early snows and a scarcity of provisions for livestock as well as the people of this small mining camp. Spring had brought new hope and the prospect of riches in the gold fields around Jacksonville.

As preparations were being made to head to the gold fields, eleven miners from California arrived in town and purchased large amounts of supplies. Although they were secretive about their mission, one Californian became quite talkative and told how their leader knew of certain landmarks, which would surely lead them to the Lost Cabin Mine. This discussion was of great interest to some of the locals, and when the Californians left town the

Figure 1-1 William Gladstone Steel *(1854-1934) – Steel's first visit to Crater Lake spawned the national park idea for the region. This thought was shared with Major C. E. Dutton, who suggested having President Cleveland reserve ten townships for a public park, as did professor Joseph LaConte. Steel is responsible for naming many of the features in the park as well as initiating much of the early scientific work. His association with Crater Lake continued throughout his life (Plate 2).*

next day, eleven Oregonians followed on their trail. A game of "hide and seek" developed between the two groups until provisions began to run low. Then for a few days, they searched for the mine and hunted game together. The lack of food caused most of the men to return to Jacksonville. Seven, however, continued the search, and it was during this time they made an amazing discovery.

Upon reaching the crest of the ridge, the three miners encountered a deep basin containing the most beautiful lake any of them had ever seen. Cliffs towering over a thousand feet high surrounded the deep, blue water of this mysterious lake that stretched out before them. Hillman was led to call it the bluest lake he had ever seen, and the others made similar comments. After considerable discussion, including a vote, they commemorated the discovery by writing "Deep Blue Lake" on a page torn from the back of a memorandum book one of the Californians carried. This was placed on a stick near Discovery Point along the rim just west of Rim Village (Plate 1). Upon returning to Jacksonville their discovery was reported, for no one knew anything about the lake the party of prospectors had found. Since gold was more important than spectacular blue lakes, the discovery was quickly forgotten.

Figure 1-2 John Wesley Hillman *– Hillman, only 21 at the time, is often given credit by Steel for the discovery of Crater Lake. The location of Discovery Point is based on a description by Hillman. (NPS image.)*

Early Visitors

Although it's not known if any of this original group ever returned to the site, other early residents did visit the lake. Nearly ten years later, on October 21, 1862, a party of mule packers and gold miners led by Chauncy Nye happened onto the lake while returning to Jacksonville from the John Day country in eastern Oregon. Their brief description and location of "Blue Lake," as they had named it, appeared in Jacksonville's *Oregon Sentinel* and became the first published report. In early August 1865, two soldiers from Fort Klamath stumbled upon Crater Lake while hunting game for a road-building party that was clearing a wagon route between the Fort and the Rogue River. Later they returned accompanied by Captain E. B. Sprague and several curious civilians to make the first descent down to the water. Again, the *Oregon Sentinel* published an account of their discovery of "Lake Majesty."

> **Box 1-1**
> **The Origin of Crater Lake**
> A Klamath Legend
>
> A long time ago, the Chief of the Below World fell in love with Loha, the beautiful daughter of a chief. He invited her to live with him in his home under a volcano (now called Mount Mazama). He promised her eternal life free of sickness, sorrow, and death. She refused. He responded by standing on top of his mountain and threatening Loha's people with a "Curse of Fire." The mountains shook, red-hot rocks hurtled down, and burning ashes fell like rain. An ocean of flame devoured the forest and flowed down the valleys until it reached the homes of Loha's people. They fled in terror to Klamath Lake for safety. Two great medicine men of the tribe went up the mountain and jumped in the fiery pit of the Chief of the Below World. Upon seeing the bravery of these wise men, the Chief of the Above World standing on Mount Shasta drove the Chief of the Below World back into his home. Then he caused the top of the mountain to fall in on him. The next morning when the sun rose, the mountain was gone, and the Curse of Fire was gone. Rain fell for many years to fill the hole left when the mountain collapsed to form Crater Lake. (Clark, 2003)

Visits became more frequent as Jacksonville residents and Fort Klamath military personnel made trips to this unusual place. Among these early visitors to Crater Lake was Jim Sutton, editor of the Jacksonville newspaper. After some investigations and consideration of how the lake was formed, he suggested that it occupied the crater of an extinct volcano. With this in mind, Sutton wrote an account of his visit in 1869, in which he referred to "Crater Lake," and thus the name came into common use. It seems that early settlers never learned of the lake's existence from the Native Americans, even though one of the two soldiers in the 1865 hunting party was married to an Indian woman.

Although the members of the Hillman Party lay claim to the discovery of Crater Lake, they were certainly not the first to know of its existence. Evidence is clear that Native Americans who resided in the region were familiar with this place. A number of their stories refer to Crater Lake and/or its origin (Box 1-1).

The Klamath Indians revered Crater Lake as home of Llao, a powerful spirit and chief over many of the lesser spirits. Because of their belief that punishment might visit those who looked upon places sacred to spirits, few of the natives had ever been to Crater Lake. This may also explain why the early settlers had not learned of the lake before. Later, although Hillman located Indians who knew about Crater Lake, they refused to discuss any details.

The person most intimately associated with Crater Lake was William Steel

(Figure 1-1). He learned of its existence while still a school boy in Kansas, supposedly from reading a newspaper article wrapped around his lunch. Upon seeing the lake for the first time in August 1885, Steel was so impressed with its beautiful wild state that he "came under conviction" to save this "wonderful spot." Over the ensuing seventeen years, his writings, lectures, and personal finances were dedicated to this cause. It is quite accurate to say that William Gladstone Steel is the father of Crater Lake National Park, for it was largely through his efforts that Congress created the nation's fifth national park in 1902.

Early Geological Research

In Steel's efforts to gain national attention for Crater Lake, he encouraged scientific exploration of this unusual geological feature. A reconnaissance of the geology was conducted by Major C. E. Dutton in 1883, which clearly established the fact that a large volcanic cone had once stood at the site. A later Dutton survey, sponsored by the United States Geological Survey (USGS), made the first detailed soundings of the lake during the summer of 1886 to produce a map showing the lake's depths (Figure 1-4). Today, only the lower portion of this major Cascade volcano remains.

Figure 1-3 Joseph S. Diller – *Diller and Patton's geological studies during the 1890s resulted in the first major scientific explanation and publication on the origin of Crater Lake. Their collapse hypothesis has proven to be correct.* (NPS image.)

Some fifteen years later, J. S. Diller and H. B. Patton conducted a detailed geological study over several seasons. Their work included collecting and analyzing rock samples, as well as a detailed study of the surface geology (Figure 1-5). With their final report, issued in 1902, the first complete geological story of Crater Lake was proposed. After Dutton established the presence of a large volcano, a popular story was developed that suggested a great explosion blew it apart. Diller and Patton's study, however, was even more exciting, for they offered a far different and unexpected explanation of what happened to the volcano that stood where Crater Lake is today. In their work, it became apparent that something caused the upper portion of the volcano to col-

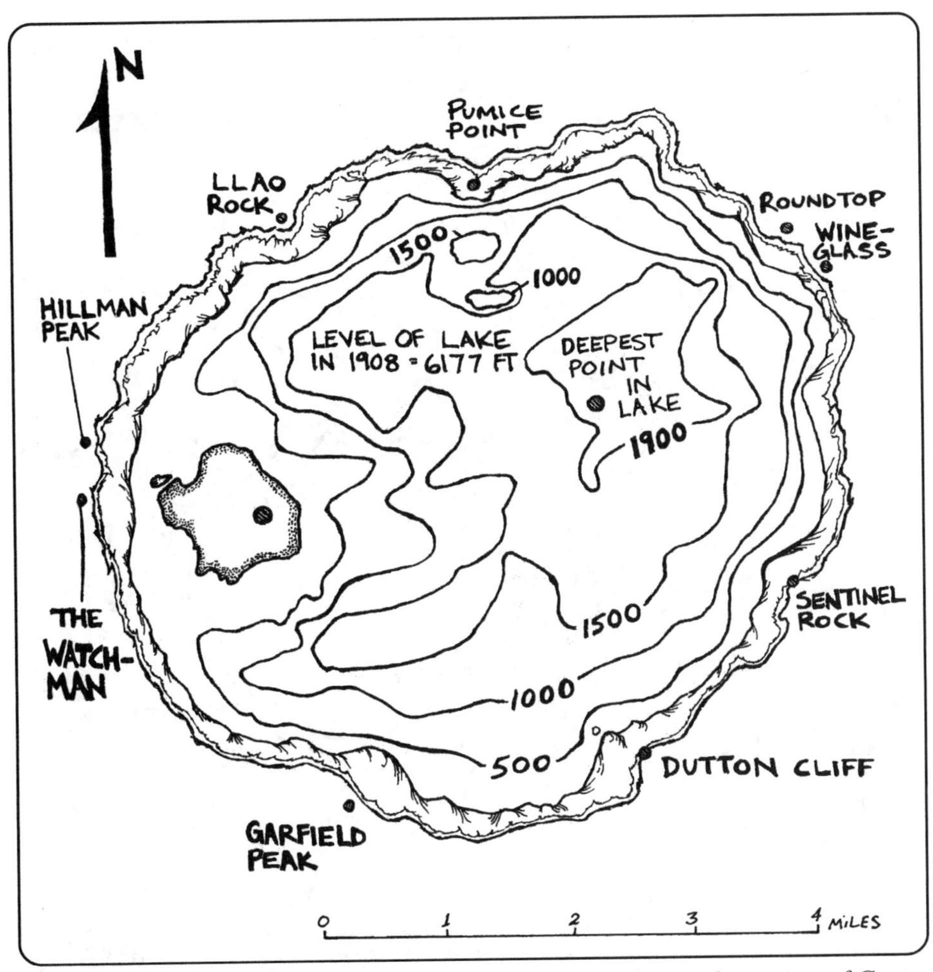

Figure 1-4 Crater Lake's Floor – *Prior to 1957 the configuration of Crater Lake's floor was primarily based on the 168 soundings from the 1886 Dutton Survey. Later, park naturalists also added some depth measurements, which are included on this map. Depth contours in feet. (After Williams, 1942.)*

lapse into the interior, forming a deep surface depression. When the remains of this volcano filled with water, Crater Lake was born.

Again, it seems the Native Americans must also be given credit for understanding how Crater Lake formed. Although their explanation is couched in terms of spirits and legends, it is essentially a description of what modern geology has found. This should not come as a surprise because it is clear, based on archeological evidence from the 1930s, that they were likely eyewitnesses to the actual event that created the basin that now holds Crater Lake.

As geological field studies were being carried out in the 1890s, a group of mountain climbers from Portland, known as the Mazamas, held their annual

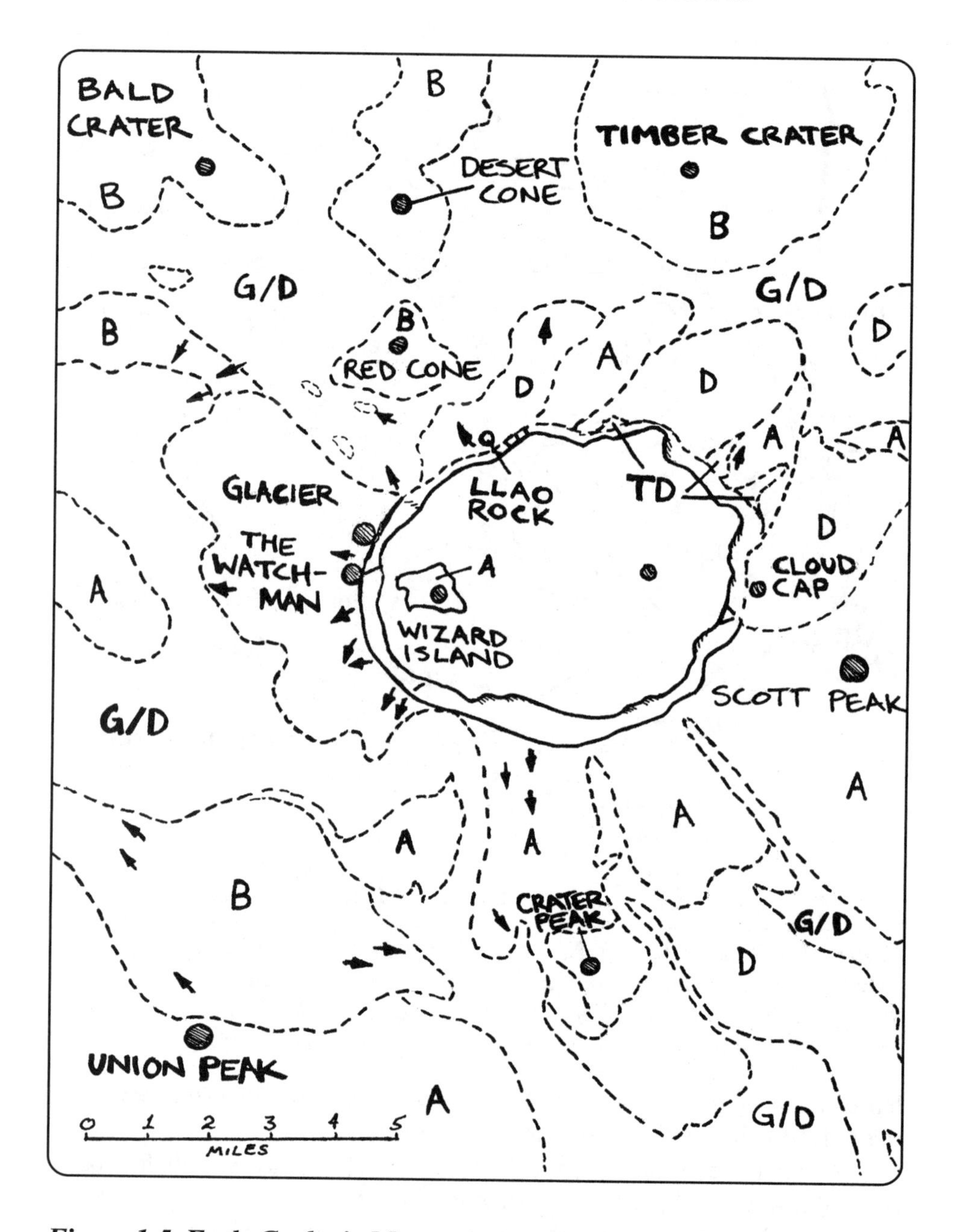

***Figure 1-5 Early Geologic Map** – This sketch is based on Diller and Patton's colored geologic map of Crater Lake National Park produced for their 1902 professional paper. It was printed on a topographic base and included the depth soundings made by the 1886 Dutton Survey. Key: G/D = Glacial Moraines and Dacite, TD = Dacite Tuff, D = Dacite, B = Basalt, A = Andesite, ⟶ = Glacial Striations, \|\| = Dikes.*

meeting at Crater Lake. This club, organized by Steel in 1894, gathered for their 1896 outing along the southern rim. Also invited were a number of prominent scientists, including Diller. Activities included nature walks and evening campfire lectures describing the natural history of the area. As a part of the official business to be transacted, Steel proposed that the ancestral volcano be named for the Mazama Club. This suggestion was favored and a new name appeared on the map in southern Oregon – Mount Mazama.

Modern Geological Research

Since its establishment as a national park, Crater Lake has been the site of extensive geological research. It is, perhaps, the best known and most studied example of a caldera in the entire world. Two researchers stand out in their contribution to a detailed understanding of the geological history of Crater Lake National Park: Howel Williams and Charles Bacon.

When Williams began field studies in the late 1930s, the accepted explanation of how the Crater Lake basin was created was still being debated. Even though Diller and Patton had offered compelling evidence for collapse – the formation of a caldera – some geologists still argued for a massive explosion. By estimating the likely volume of the original volcano and carefully measuring the amount of ejected material produced during the final eruptions, Williams and others were able to demonstrate that Mazama had collapsed. His 1942 publication, *The Geology of Crater Lake National Park, Oregon*, detailed the sequence of events that occurred from the building of Mount Mazama through the climactic eruptions and final destruction as the summit floundered into the partially emptied magma chamber below.

***Figure 1-6 Howel Williams** – In the late 1930s Williams' field studies in the park established the fundamental geology story of Crater Lake, published in his classic monograph, which is still accepted today.* (NPS image.)

Volcanology progressed rapidly during the middle portion of the century. By the late 1970s, the understanding of how volcanoes behave had increased dramatically. This was the situation when United States Geological Survey Geologist Charles Bacon began work at Crater Lake in 1980. He conducted field studies in the park each summer for ten years and continues to visit periodically. Building on Williams' work and using modern analytical techniques, Bacon has further detailed the geological history of Mazama's construction, climatic eruptive period, collapse to form the caldera, and subsequent volcanic activity that has occurred at Crater Lake. Much of the details involving the geology of Crater Lake National Park presented in this book are based on Bacon's research, interpretation, and publications.

Figure 1-7 Charles R. Bacon – *Bacon's field studies were conducted during the 1980s and 90s and added important details to the interpretation of Crater Lake's geological history.* *(NPS image.)*

Crater Lake National Park

President Theodore Roosevelt had been in office only a few months when he signed the legislation establishing Crater Lake National Park in south central Oregon. The date was May 22, 1902. Several bills had been introduced previously to create a national park in this remote area of the country. In fact, efforts date back to at least 1886, when President Grover Cleveland withdrew ten townships around Crater Lake to form a reserve in anticipation of Congress passing legislation to establish a national park. His action followed on the heels of pending federal legislation that would create a "public park" around Crater Lake.

The creation of Oregon's only national park was the culmination of more than sixteen years of maneuvering at various levels of government. President Grover Cleveland's action created a reservation around Crater Lake, temporarily protecting it from settlement or sale. On the first map of Klamath County drafted in 1889, the words "National Park" appeared to distinguish the reserve from surrounding land. Bills to create Crater Lake National Park were introduced in both the United States Senate and House of Representatives during the

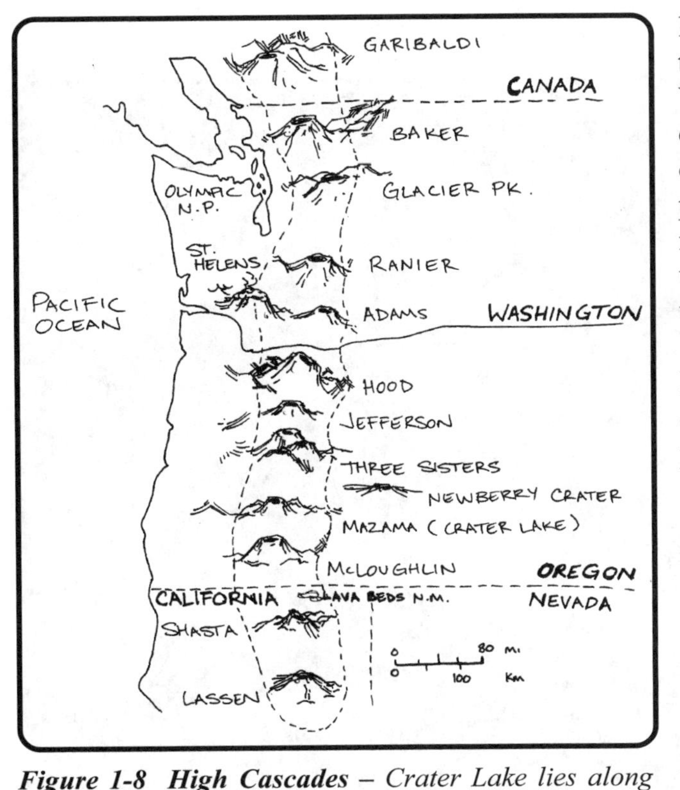

Figure 1-8 High Cascades *– Crater Lake lies along the High Cascades, the most active volcanic region in the contiguous United States.*

late 1880s and throughout the 1890s. While most died in committee, the idea continued to gain support. Finally, after many years, Steel's unrelenting efforts were rewarded in 1902 when Crater Lake became this country's fifth national park.

Additional Reading

Clark, E. E., 2003, *Indian Legends of the Pacific Northwest*, University of California Press, ISBN 0520239261, 240 p. A collection of Native American myths, legends, and tales from the Pacific Northwest.

Harmon, R., 2002, *Crater Lake National Park – A History*, Oregon University Press, ISBN 0-87071-537-2, 280 p. A detailed history of Crater Lake National Park focusing on administrations.

Mark, S. and Gnass, J., 1996, *Crater Lake: The Continuing Story (in pictures)*, KC Publications, ISBN 0-88714-109-9, 50 p. Primarily colored photographs with captions.

Warfield, R., Juillerat, L., and Smith, L., 1982, *Crater Lake: The Story Behind the Scenery*, KC Publications, ISBN 0-916122-79-4, 50 p. Covers all natural aspects of Crater Lake with many colored photographs.

Chapter 2
Cones, Craters, and Calderas

Perhaps the most common image that comes to mind with the term volcanism is red-hot lava flowing down the flanks of an erupting volcano. After all, such pictures are frequently shown to illustrate volcanic activity – usually from some currently erupting vent. There are, however, many other types of material ejected from volcanoes, including hot gases, lava too thick to flow, and solid materials, which are ash, bombs, and blocks. A two-fold classification system has been developed for volcanic rocks. The liquid is called lava, whereas materials ejected as solid particles are known as pyroclastics ("fire broken" rocks).

Lavas

Because lava is hot enough to flow when temperatures are between 850°C and 1,150°C (1,300°F and 2,100°F), it must cool before solidifying into rock. Depending on its temperature at the time of eruption and the cooling conditions encountered, various types of lava will form. Pahoehoe (Pa-hoy-hoy, a Hawaiian term) is the most fluid, and while cooling takes place, a crust may form on the surface of the flow. As the flowing lava below carries this partially solidified skin along, it may be deformed or broken. But as the flow slows down, the surface piles up to form ropy-like patterns (Figure 2-1) while rapid movement of the

Figure 2-1 Ropy Pahoehoe – *This form of lava is usually associated with basaltic compositions and is therefore not common in the andesites or dacites found at Crater Lake. This pahoehoe flow occurred at Craters of the Moon National Monument in southern Idaho. (Note knife for scale.)*

underlying lava causes fracturing into slabs that tend to float on the surface. Under certain conditions the hardened crust may be strong enough to permit the lava to flow away completely. This can leave a tunnel-like feature known as a lava tube (Figure 2-2).

Figure 2-2 Lava Tube *– This large lava tube, formed in a basaltic flow at Craters of the Moon National Monument, is partially collapsed.*

In cooler flows, the skin is thicker but much less plastic. Thus, when movement of the underlying lava occurs, this rigid surface breaks into blocks (Figure 2-3). This blocky lava is often given the strange name of aa (ah-ah, another Hawaiian term). While actually forming, aa flows appear to be slowly moving piles of clinkers, which may provide an occasional view of the hot lava inside.

Pyroclastics

Pyroclastic ejecta takes on many more forms than fluid lava. This eruptive material is usually pieces of crystallized liquid but may also be liquid that cools so rapidly that it forms solid glass, or even nonvolcanic rocks caught up in an eruption. Geologists use the term tephra for volcanic rocks formed by explosions from a volcanic vent. The classification of tephra is based on the size of individual particles, as listed in Table 2-1. The smallest category, ash, may be small indeed and is often transported great distances from the place where it was erupted (Figure 2-4). With increased size, ash particles are transported shorter distances, with the largest pieces falling out near the site of the eruption. While the term ash represents a particle size, it is also used in a way to describe how volcanic material is deposited.

Figure 2-3 aa Lava *– This form of lava can be found at some locations in the park. One of the best examples makes up the flows around the base of Wizard Island. (Note the knife for scale.)*

Table 2-1 Classification of Tephra

Name	Size	Description
Ash	Less than 2 mm (3/64 in.)	Consists of glassy particles easily transported by wind.
Lapelli	2 mm to 64 mm	Often spheroidal in shape but may be elongated. Developed from liquid ejecta or by the accretion of smaller particles. Term means "little stones."
Bombs	Larger than 64 mm (2 1/2 in.)	Clots of magma partly or entirely plastic upon eruption. The shape is the result of initial fluidity, rate of cooling, time in flight, and other factors. There are several types of bombs.
Blocks	Larger than 64 mm	Erupted fragments of solid rock, usually angular and equidimensional but may be slabby or platey.

When ash-sized material is carried through the air and falls to the surface forming deposits, it is called ash-fall. During violent eruptions ash, in a mixture of gas and other pyroclastic material, may be carried along the surface in an expanding cloud of debris. Rocks or loose materials resulting when these volcanic ejecta

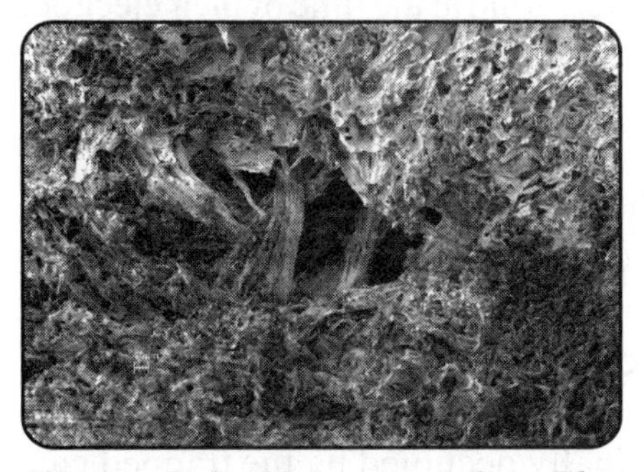

***Figure 2-4 Mazama Ash** – This scanning electron microscope image of Mazama's climactic eruption ash shows the glassy and porous nature of pumice. It was deposited during the Ring-Vent Phase and blankets most of the surface around the caldera.* *(courtesy image.)*

***Figure 2-5 Lapilli** – This "popcorn" pumice deposited along the East Rim Drive near Cloud Cap Junction is coarse lapilli from the Red Cloud eruption. The hammer head is 18 cm (7 in.) long.*

Figure 2-6 Cinders and Pumice Blocks *Cinders have the appearance of tiny "clinkers" and are usually dark colored due to their chemical composition (Fe and Mg content). Pumice may occur as particles of any size, ash to blocks, looks spongy, and is usually light colored with low density.*

come to rest are termed ash-flow deposits. They are commonly associated with the formation of calderas, and some of the rocks found on the flanks of Mount Mazama represent deposits from the caldera-forming event that created the Crater Lake caldera (more in Chapter 6).

Although lapilli is not as abundant as the ash-size particles, it is still common in volcanic areas (Figure 2-5). The marble-size particles associated with cinder cones and other explosive eruptions fall into this class. Perhaps the most spectacular form of tephra is blocks and bombs (Figure 2-7). While blocks are angular-shaped fragments and were most likely solid at the time of their ejection, bombs often have a more rounded appearance. Highly streamlined shapes suggest that bombs were molten when ejected and achieved their shape while moving through the air.

Two other terms are also used in the discussion of tephra: pumice and cinders (Figure 2-6). When silicate liquid experiences a sudden loss of pressure, dissolved volatiles (typically H_2O and CO_2 gas) form as bubbles. If the liquid was viscous (thick), this inhibits the escape of gas that was dissolved in the fluid rock. The result is a glassy rock with a sponge-like structure. Most of the volume is taken up by empty space, originally occupied by the trapped gas, separated by thin glass membranes. This is pumice. Cinders, or scoria as they are commonly called, have much the same history. The basic difference is in the amount of empty space present and the chemical composition of the rock. Since the original material that forms into scoria is mostly fluid, trapped gases escape more readily and thus there is less void space and more rock material. Also, as noted in Table 2-2, cinders are considered to be lapilli, while pumice has no particle size connotation. In addition, generally, cinders are associated with erup-

Figure 2-7 Volcanic Bombs *– Bombs are formed in many sizes, although the smaller ones display the most streamlined shapes. (A) The streamlined shapes of two bombs, which were fluid when erupted, can be readily identified. (B) A large spherical bomb. (C) The same bomb as shown in B, illustrating a concentric internal structure that develops upon cooling. The crust has a fine texture due to rapid cooling and becomes coarse toward the center, where cooling was slower. (D) This breadcrust bomb developed when expanding gases caused swelling and cracking of the hardened surface as the inside cooled and shrank.*

tions of a basaltic to andesitic composition, while pumice is composed of dacite or rhyolite.

Box 2-1
Volcanic Gases

Analysis of gases found in the dacite pumice from the 1902 eruption of Mount Pelée. While this may not be typical of all volcanoes, it was an explosive eruption.

Gas	Percent
Water	71%
Chlorine	11%
Carbon dioxide	7%
Fluoride	4%
Sulfur	3%
Nitrogen/Argon	2%
Carbon monoxide	1%
Hydrogen	< 1%

Source: Williams and McBirney, 1979

Gases

Due to the high pressure in magma, most gases are in solution, something like in a carbonated drink. When the pressure is released, during an eruption, some gases return to the gas state of matter and escape – the primary force in explosive eruptions. When water changes to a gas state, its volume is increased a thousand times, instantly! As expected, obtaining good data on the composition of volcanic gases during an eruption is difficult, and most come from "quiet" eruptions like those in Hawaii that are basaltic in composition. Collecting gas samples during an explosive eruption is much more challenging. However, this can be done with remote instruments now, and volatile amounts can be measured in gases trapped in glass. From those analyzed so far, it appears there are a large range of compositions. Water is certainly an important component and in some cases the vast preponderance of dissolved gases escaping during an eruption. Others, however, may also be present if not abundant, including carbon dioxide, sulfur dioxide, hydrogen sulfide, nitrogen, and others (Box 2-1). In images of massive eruptions showing huge billowing white clouds, water vapor is the principle component creating the clouds.

Chemical Composition

Still another factor in understanding volcanic rocks, including those at Crater Lake, is their chemical composition. Volcanic rocks are a type of igneous rock formed by the freezing of hot, liquid material called magma. Since cooling of magma occurs quickly when it reaches the Earth's surface (where it is called lava), volcanic rocks tend to form extremely small crystals. These are usually too small to see without magnification, although larger crystals, big enough to be easily visible, are often present in lava. These larger crystals were formed prior to an eruption and are known as phynocrysts. Volcanic rocks are composed of many elements, but only a few are abundant: oxygen, silicon, iron, aluminum,

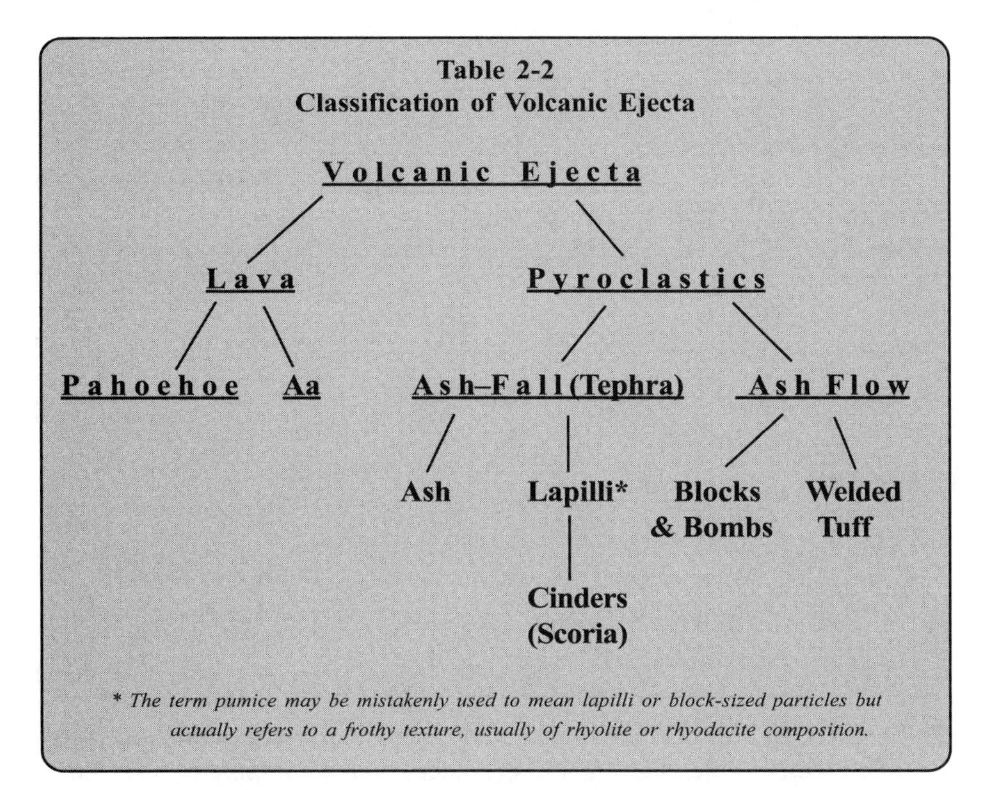

calcium, magnesium, sodium, and potassium. Moreover, there is usually far more oxygen and silicon present than all the others combined. As a result, these two elements, expressed as one silicon atom joined with two oxygen atoms (SiO_2), are used to describe the chemical composition of igneous rocks. Table 2-3 illustrates a simple classification of volcanic rocks and some of the properties associated with certain compositions. The values shown are approximate and, of course, the entire classification scheme is artificial for there is a continuous composition range in volcanic rocks from as low as 35% to as high as 75% SiO_2.

Cones

When asked to imagine a volcano, most people think of the large picturesque volcanoes commonly seen on calendars and postcards. Although these beautiful symmetrical cones may represent all volcanoes, at least in people's minds, they are really the exception; few take on this pleasing shape. What, then, does a typical volcano look like? To answer this question, we must look at different types of volcanoes, formed in different ways and from different material. Let's start by grouping these into four general classes as noted in Table 2-4.

Two of the best examples of composite cones are Mount Mayon in the

Table 2-3
Volcanic Rocks and Some of Their Properties

Name	SiO_2	Color	Amount of Gas	Nature of Lava
Rhyolite	**70%** (or higher)	**Light** (buff or tan)	**High**	**Very thick – seldom flows** **Fe & Mg poor (< 3%)**
Dacite	**65%**			
Andesite	**60%**	**Medium**	**Medium**	**Flows slowly**
Basaltic-Andesite	**55%**			
Basalt	**50%** (or less)	**Dark** (black or green)	**Low**	**Fluid – flows easily** **Fe & Mg rich (up to 5%)**

Philippines and Mount Fuji in Japan – Fuji is one of the largest on Earth. Both are often seen on calendars and travel posters. Many composite cones, however, do not possess the pleasing symmetry illustrated by these ideal examples. The construction of composite volcanoes involves many factors, and they have a complex eruption history. Two of the major contributing influences are the composition of the volcanic materials involved and the manner of growth.

Erupted volcanic materials range in composition from basalt to rhyolite, but most composite cones are built of andesite (Table 2-3). These lavas with a higher silica content and lower temperatures tend to be more viscous and do not flow as far. This results in building a steeper cone. Often a composite cone's composition changes over time, becoming more silica-rich, which further adds to this steepening effect. Also, as ejected volcanic material becomes richer in SiO_2, the amount of gas involved increases, and eruptions become more explosive in nature. Although the bulk of volcanic material making up a composite cone is usually lava, pyroclastics may also be important in their construction. In examining the volcanic materials that compose composite cones, it is not unusual to find lava flows interspersed with pyroclastic layers. The shape and slope of such cones are influenced by the relative proportions of lava and pyroclastic materials produced and deposited in a more or less alternately layered nature. When a volcano erupts coarse pyroclastic debris, it will accumulate near a volcano's vent, creating a steeper slope as compared to lava flows that tend to flow further from the vent (Figure 2-8).

Table 2-4
Types of Volcanoes

Name	Size: Height	Size: Diameter	Type of Material
Composite	**Up to 3,600 m** (12,000 ft)	**Up to 40 km** (25 mi)	**Basalt to Rhyolite** major portions are andesite & rhyolite
Shield	**Up to 9,000 m** (30,000 ft)*	**Up to 160 km**	**Basalt** (primarily)
Cinder **(or Scoria)**	**Up to 350 m** (1,200 ft)	**Up to 1.6 km** (1 mi)	**Basalt to Andesite**
Miscellaneous			
Domes	**About the same as cinder cones**		**"Fluid" Dacite/Rhyolite**
Maar	**Less than cinder cones** (usually formed in or with water)		**Pyroclastic Dacite/ Rhyolite**

* The Island of Hawaii, at about 9,000 m above the ocean floor, represents the tallest mountain in the world.

Another significant factor influencing the shape of very large composite cones is the location and number of vents. Such features tend to have a long history of activity that may result in a number of active vents developing over time. Those ideally shaped symmetrical volcanoes noted above develop when eruptions are primarily from a single central vent area. If activity moves from one location to another on the volcano, the overall shape of the composite cone will be much modified and thus less symmetrical. When multiple vents occur, the overall volcano is constructed by a series of coalescing structures – these are sometimes called compound cones. This apparently was the nature of Mount Mazama, the composite cone that was originally built on the site where Crater Lake is located today.

When the volcanic material is mostly basaltic lava flows, which are quite

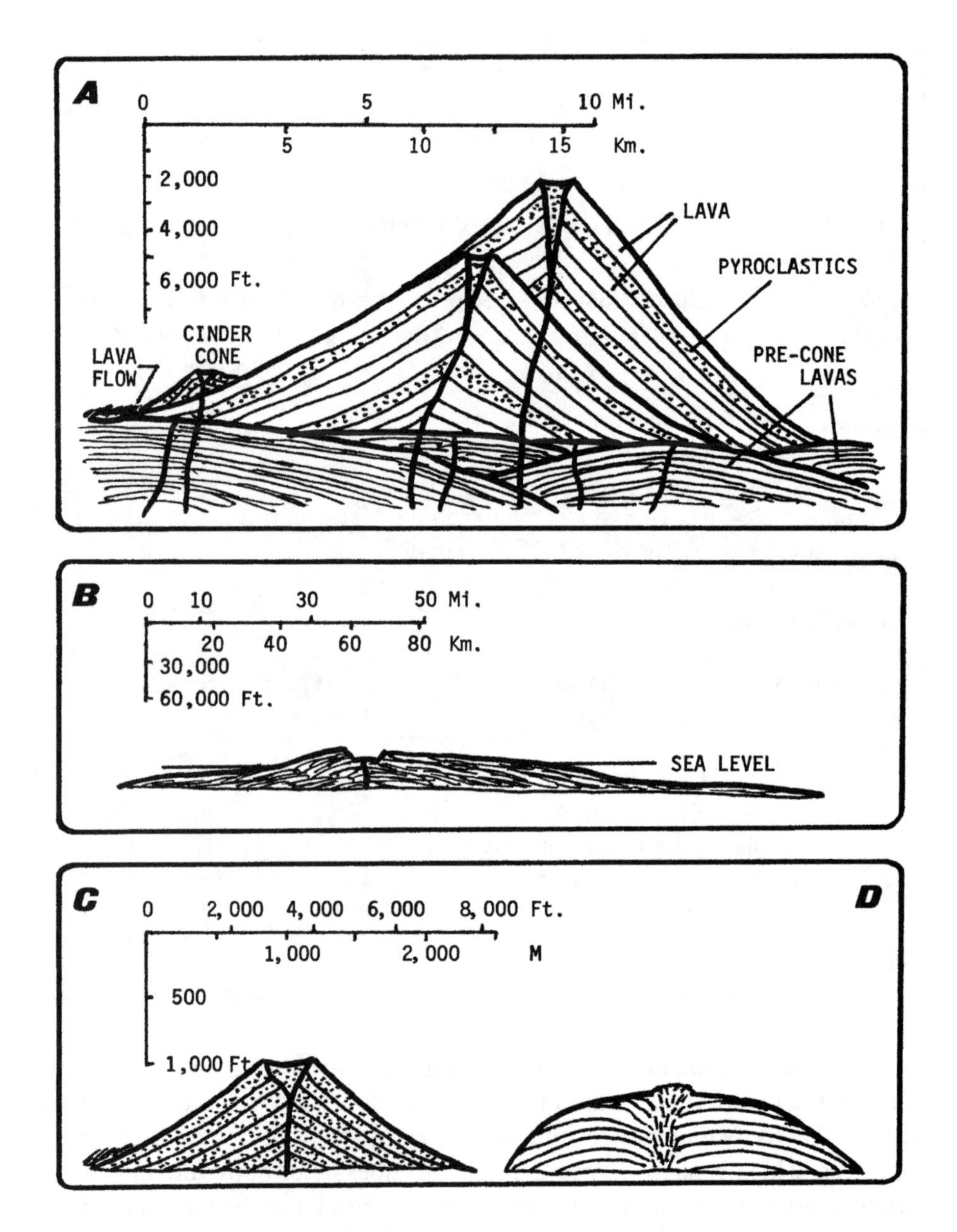

Figure 2-8 Major Types of Volcanoes *– Schematic cross sections of the four types of volcanic cones found at Crater Lake National Park. (A) Composite cone showing a complex constructional history – Mount Mazama. (B) Large shield cone similar to the Hawaiian Islands – Union Peak is a small version. (C) Cinder cone with associated basalt lava flow at lower left – Wizard Island. (D) Dacite dome – several in eastern area of park. (Notice the horizontal and vertical scales for comparison of sizes and that the scales are dramatically different for each sketch.)*

fluid, a low mound-shaped cone develops. These are generally called shield volcanoes but may be given other names, such as lava cones. Hundreds or thousands of individual flows, each only a few feet thick, are responsible for building shield cones. The largest volcanoes on Earth, indeed in the entire solar system, are of this type. Although some pyroclastic material, in the form of scoria, is often erupted during the construction of a shield cone, the amount is insignificant when compared to that of basaltic lava flows. The massive piles of volcanic material in the Pacific Ocean, known as the Hawaiian Islands, are classic examples of this volcanic cone type. To provide a sense of size relationship between large shield cones and large composite cones, the big island of Hawaii has a volume about 300 times that of Japan's Mount Fuji.

Moving on to a third type of cone and examining volcanoes that are composed primarily of pyroclastic ejecta, usually basaltic or basaltic-andesite in composition, we are looking at cinder cones (Table 2-4). These small vents are likely the most abundant type in the world, for they form by the hundreds in any given volcanic region. For example, there are over 400 cinder cones in the Oregon section of the Cascade Range. Cinder cones (also known as scoria cones) seem to develop in two stages: (1) an initial explosive period, which exhausts most of the gas stored in the erupting lava, followed by (2) quiet lava flows in a later eruptive stage. Although the lavas may be abundant in some cases, the greater portion of these volcanic cones appear to be composed of pyroclastic ejecta. And while all sizes of tephra are present, lapilli-size cinders seem to be dominant, with a good representation of small blocks and bombs. The late lava stage(s) seldom erupt from the upper portion of the cone but commonly break out at the base only to flow away. Cinder cones have extremely short life spans when compared to large volcanoes. From the time of the initial eruption to its final sign of life, a cinder cone may be active for only a few months to a few years. They seldom, if ever, renew eruptions following a period of dormancy after activity stops.

The discussion above describes the three most common types of volcanoes: composite, shield, and cinder cones. There are, however, many others that are less abundant but equally interesting. One of these, usually called a dome (or rhyolite dome), is not so much a volcanic cone as it is a "blister" formed by extremely thick lava (due to high SiO_2 content). Upon reaching the Earth's surface, this magma is too thick to flow, so it erupts by squeezing up and freezing to form a mound – something like toothpaste oozing out of a tube. After the explosive eruptions of Mount St. Helens in the early 1980s subsided, an active dome formed in its crater (Figure 2-9).

Volcanic cones found in the Crater Lake region are confined to these four types. It should be noted, however, that large amounts of magma reach the Earth's surface through fissures that do not construct volcanic cones at all. The great mass of basaltic lava making up the Columbia Plateau and Snake River Plains were formed from such fissure flows (Chapter 3). Much of the magma rising along the Cascade Range in recent geologic time, however, tends to produce volcanic cones rather than massive lava flows.

***Figure 2-9 Resurgent Dome** – A silica-rich dome was built and now fills the crater at Mount St. Helens. The dome can be seen in the center of this image venting steam, which indicates the volcano is still active. Most of the volcanic activity periodically builds up the dome which then collapses.*

Craters

Another common impression shared about volcanoes is the way that eruptions occur. It is often assumed that volcanic material is ejected from the top or main vent portion of an active volcano. Volcanoes with a depression in their summit, or crater as it is properly called, do produce large amounts of material, but eruptions may also occur from other locations on the cone. Lava may break out on the sides of a composite or shield cone to produce flank eruptions. Sometimes these flank vents result in the construction of small features called parasitic cones. With changing conditions in a volcano's interior, the active area on the surface may also shift around in the summit region of a large composite volcano. This might build a number of overlapping cones with separate vents, as appears to be the case for Mount Mazama. Mount Scott (Plate 5) and Hillman Peak in the park are thought to represent two of possibly a number of summit vents. Thus, shield cones and large composite cones may have several craters, with activity in one or another at different times. It must be noted, though, that such vents on composite cones are rather small, usually less than 1 km (~0.6 mi) across.

Calderas

Some volcanoes develop a large depression in their summit area. Such features are known as calderas and result from some type of collapse that produces a more or less circular basin. Large calderas are the result of an exceptionally violent volcanic event, far beyond the normal activity typical of a crater eruption. The word caldera is Portuguese and was originally used by natives of the Canary Islands. It was first used as a geological term in the early 1900s and introduced in America by Dutton about the time he was conducting field studies at Crater Lake. While the term caldera has been used in numerous ways since it first appeared in geological literature, its modern meaning is more narrowly defined to refer to a volcanic feature.

The idea of a large collapse feature caused by a major volcanic eruption dates back to the late 1800s with a description of the 1620 B.C. eruption of Santorini in the Aegean Sea. A French geologist concluded that part of the original volcano that seemed to be missing had sunk below the water level. Shortly after the turn of the century, in 1909, a volcanic feature in Scotland was shown to have formed by collapse following the eruption of a great volume of pyroclastic material. Further advances in understanding the mechanism of caldera formation followed field studies in the western United States in the mid-1900s. In the early 1960s two USGS geologists found that these features were clearly linked to massive pyroclastic deposits. A sequence of well-defined phases were identified that helped explain the development of large calderas. These follow in a general pattern: volcanic activity associated with uplift of the area, eruption of pyroclastic material, collapse to form a depression (a caldera), post-collapse eruptions (often lava flows), and resurgence (uplift) of the area involved.

Some twenty years earlier we find a familiar name from geological work at Crater Lake – Howel Williams. In a major contribution to understanding calderas, he provides this definition from his 1941 publication, *Calderas and Their Origins,* ". . . calderas are large volcanic depressions, more or less circular or cirque-like in form, the diameters of which are many times greater than those of the included vent or vents, no matter what the steepness of the walls or form of the floor." Williams goes on to say "calderas are almost invariably large volcanic basins produced by engulfment." This same work attempted to provide a classification system for calderas.

Williams broadened his concept of calderas and refined his classification in the popular textbook, *Volcanology,* published in 1979. He recognized seven types of collapse volcanic features, three of which are associated with explosive eruptions producing pyroclastic debris. The Crater Lake caldera fits into Williams' Krakatoan type, based on the nature of the famous 1883 eruption of that

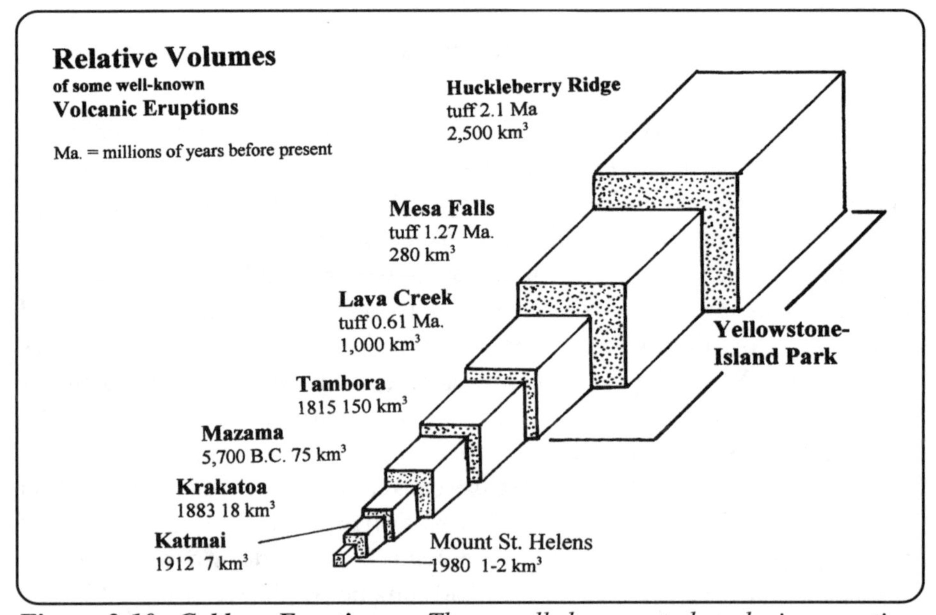

***Figure 2-10 Caldera Eruptions** – These well-documented explosive eruptions have resulted in the creation of a caldera. There appears to be a direct relationship between the amount of ejecta produced and the size of the caldera formed.(After R. Smith and L. Braile, 1984.)*

volcano located in the Sunda Straits between Java and Sumatra. This type is "formed by the foundering of the tops of large composite volcanoes following explosive eruptions of siliceous pumice from one or more vents . . . The volume of ejecta is usually much less than 100 km^3" (Williams and McBirney, p. 208).

According to Williams, one major element of a Krakatoan type event is a Plinian eruption. Such a violent volcanic phenomena appears as a massive plume exploding from the vent of a volcano. It is driven by pressure as gas is released from solution and may last for hours or even days. The massive eruption that occurred on May 20, 1980, at Mount St. Helens represents a recent Plinian eruption (Figure 6-4). Magma is converted to pyroclastic ejecta materials so rapidly that its source, the so-called magma chamber, is partially emptied. This allows the surface above the chamber to settle, forming a caldera. A detailed account of how the Crater Lake caldera developed will provide more on Plinian columns and caldera formation in Chapter 6.

Big calderas are among the largest volcanic structures on Earth and eruptions they produce may be the most energetic natural events that occur on Earth, comparable to an asteroid impact. Until the mid-1900s, most older large calderas were not recognized because erosion had altered their obvious surface expression. With a better understanding of the nature of ash flows and their relation to caldera formation, many large calderas have been identified. Calderas have a

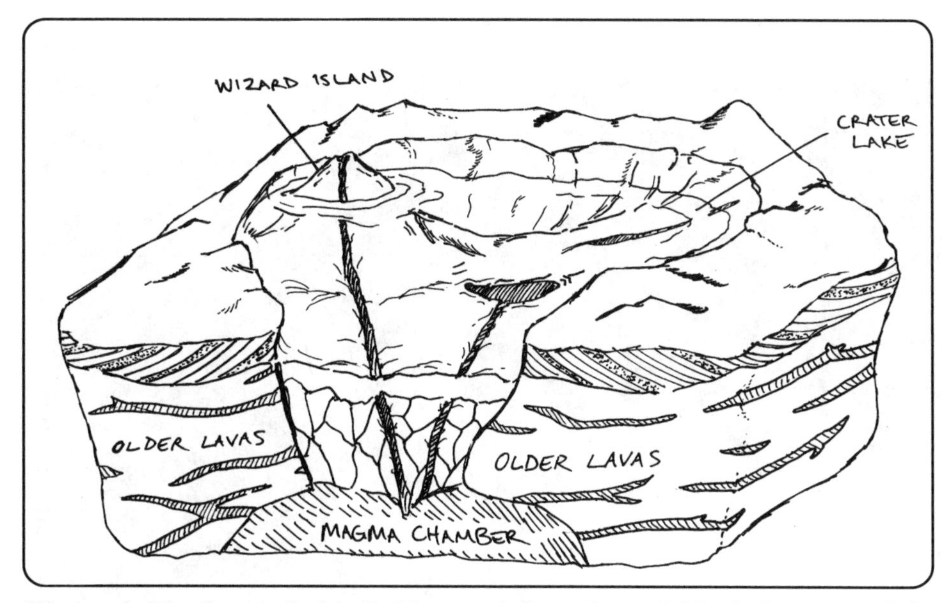

***Figure 2-11 Crater Lake Caldera** – A hypothetical block diagram of the Crater Lake caldera, illustrating the collapsed area above the magma chamber. Wizard Island's structure, the nature of the caldera floor, and the collapsed section below are only suggestive in this sketch.*

huge range of sizes, from a few kilometers in diameter to several tens of kilometers (Figure 2-10). When considering sizes of calderas on the Earth, the Crater Lake caldera is at the small end of the spectrum.

During the late 1930s Williams clearly demonstrated that such an event occurred at Crater Lake National Park when he studied its geology. Although the concept of calderas didn't originate with Williams, much of his field research during a long career established the nature of these fascinating volcanic features. His study found Crater Lake to be one of the world's classic examples of a caldera. There are calderas that are much larger, and some smaller, than Crater Lake, but few are as well preserved or illustrate the formation processes better (Figure 2-11). In recent publications, features similar in nature and size to Crater Lake are frequently called Crater Lake type calderas.

Additional Reading

Francis, P., 1993, *Volcanoes - A Planetary Perspective*, Oxford University Press, ISBN 0-19-84033-7, 443 pages. A comprehensive review of all volcanic phenomenon with a worldwide (and beyond) emphasis.

Williams, H., and McBirney, A., 1979, *Volcanology*, Freeman, Cooper & Co., ISBN 0-87735-321-2, 397 pages. This is a detailed textbook, coauthored by Howel Williams, covering all aspects of volcanology.

Chapter 3
Before Mount Mazama

The life span of Mount Mazama, the High Cascade volcanic complex that stood on the site Crater Lake now occupies, was relatively short as geological time is reckoned. It was built in less than half a million years, and its construction is the subject of Chapter 4. The geological history of Oregon and the Pacific Northwest, however, is much longer – it stretches back more than 200 Ma (Ma = million years). With such long periods of time involved, geologists use a calendar-like chart to help organize important events. This device, called the Geological Time Scale, is organized into major blocks of time that are subsequently divided into smaller lengths of time, known as periods and epochs. Noting Table 3-1, the geological history of the northwestern United States fits into the two youngest eras, the Mesozoic and Cenozoic.

A New Model

In recent decades an amazing hypothesis has been developed to explain many of the Earth's features that have long puzzled geologists, geophysicists, and oceanographers. One question involves why volcanoes are found in only certain places on the Earth's surface, usually in linear patterns. A related question centers on the location of earthquakes. Although earthquakes are relatively uncommon in the Cascades, this mountain system does represent a major volcanic chain. This new concept is known as plate tectonics (Box 3-1). It suggests that the Earth's

Figure 3-1 Mount Thielsen *– This 2,800 m (9,000 ft) peak, some 20 km (13 mi) north of Crater Lake, is known as the "Lightning Rod of the Cascades." It represents an older series of Cascade volcanoes that had ceased activity before the current large peaks were constructed on the underlaying basaltic platform. Thielsen's massive vent plug and associated dikes have been extensively eroded by Pleistocene glaciers to produce the "Matterhorn-like" peak seen here.*

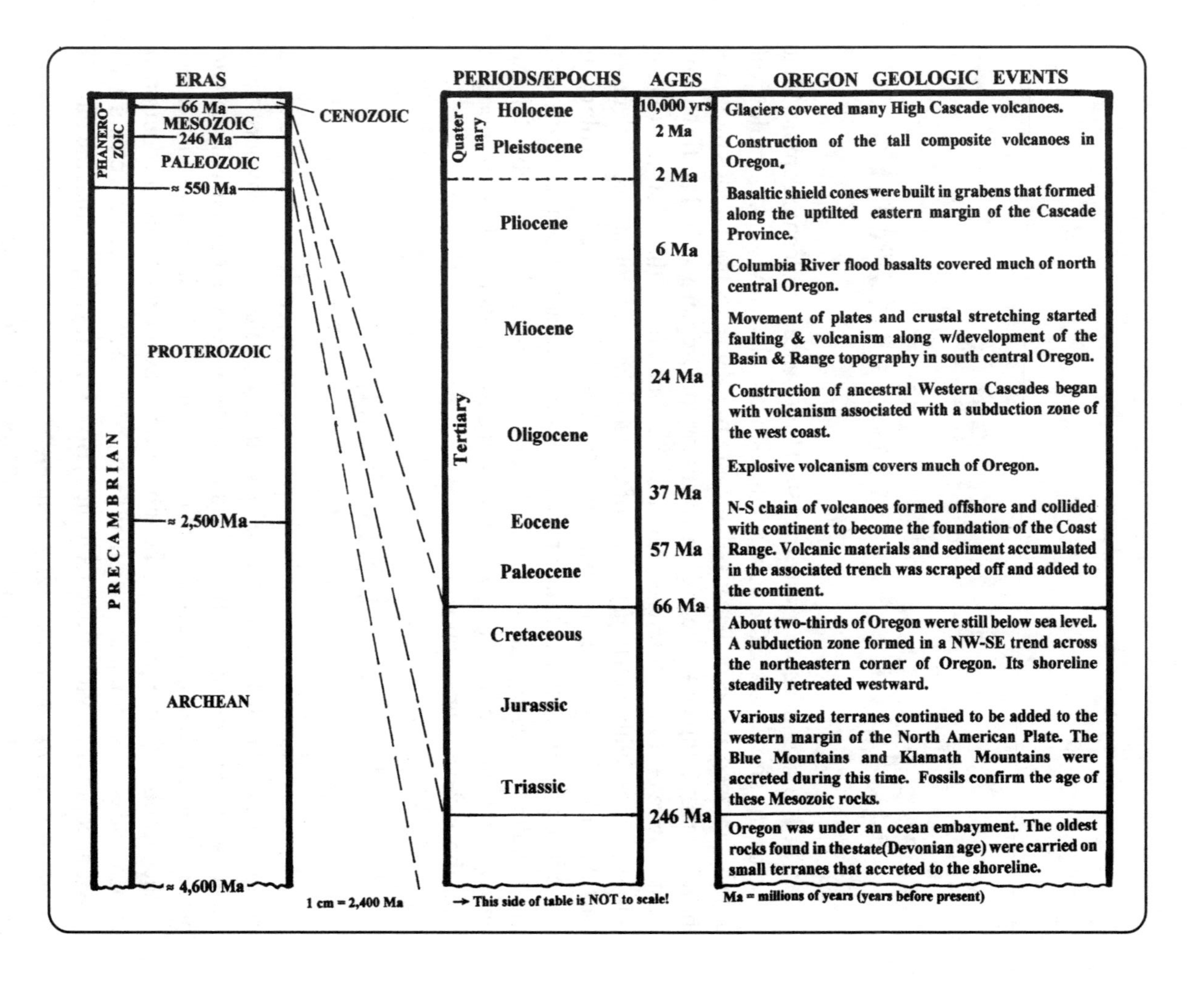

Table 3-1 – Geological Time Scale for Oregon

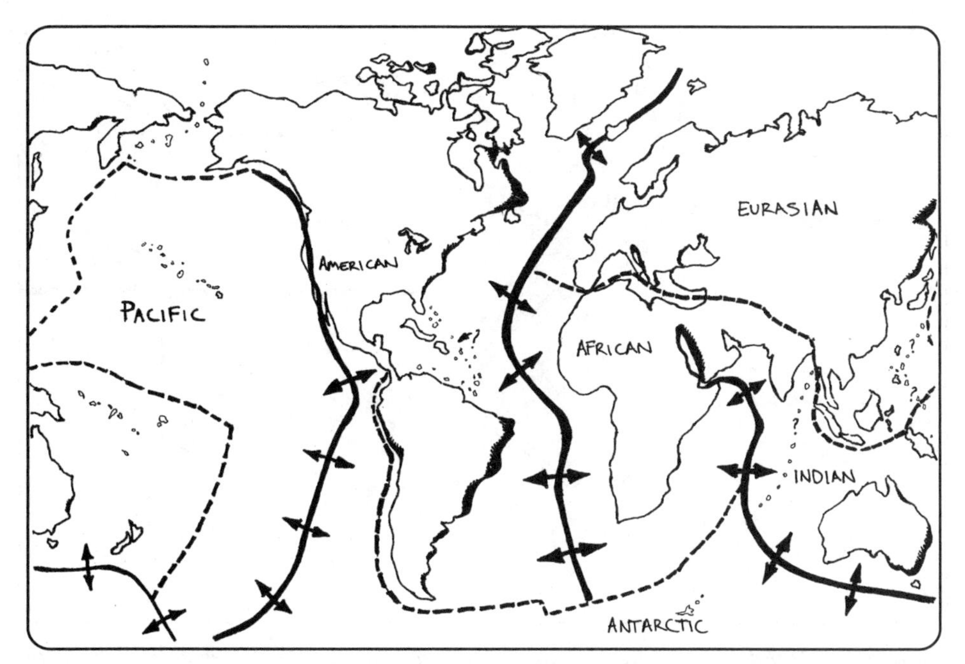

Figure 3-2 Major Plates – *The Earth's surface is composed of six major plates separated by mid-ocean ridges (<–>) and trenches (subduction zones, - - -). New crust wells up from the mantle along the ridges, moves away, and eventually slides back beneath the crust at subduction zones.*

surface is composed of several continental-sized segments, called plates, which are moving with respect to one another (Figure 3-2). Such movement may involve the separation, collision, or sliding of one plate with respect to another along their common boundary. Most of the interesting and dramatic geological activities associated with the Earth's surface, such as volcanoes and earthquakes, take place along these plate boundaries.

There seems to be a strong relationship between two colliding plates, and they may produce a feature known as a subduction zone. When the two plates consist of one below an ocean basin and the other a continental mass, this interesting feature develops.

Box 3-1
Plate Tectonics

It has been over a hundred years since the idea that continents move over the Earth's surface was suggested. Then in the early 1900s, much circumstantial evidence from a number of disciplines was presented to support the concept of "continental drift," as it was called then. However, few accepted the idea until more detailed evidence was discovered, mainly from exploration of ocean floors, during the late 1950s and 1960s. By the 1970s, most of the Earth Science community was "on board," and plate tectonics has since dominated much of geological research.

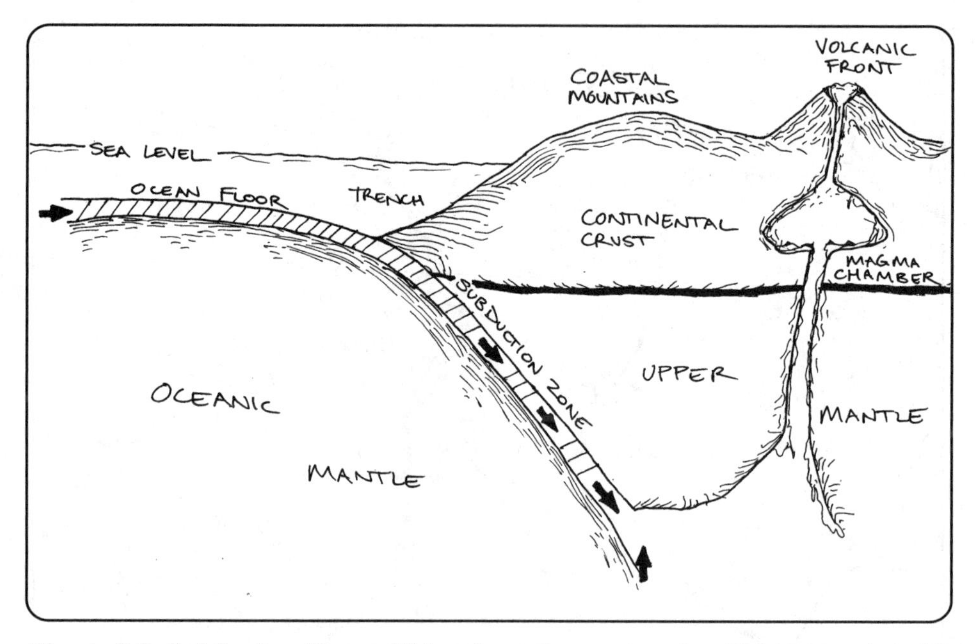

Figure 3-3 Subduction Zone *– This schematic cross section illustrates a basaltic ocean floor as it moves beneath the continental margin. It encounters a region of greatly incresed temperature and pressure that may produce magma which eventually rises to the surface as volcanoes or fissure flows.*

The ocean plate's composition is basalt, a dense rock, while the continental plate is more complex but made of much less dense rock. This results in the heavy ocean plate sliding beneath the continental plate (subducting) to produce a subduction zone. Good evidence suggests that volcanism occurs when magma is generated by the down-going oceanic slab (Figure 3-3). As temperature and pressure increase in this basaltic slab and the overlying continental mantle, melting occurs. The liquid produced should migrate upward as it is less dense than the surrounding rock. There is, however, some uncertainty in determining if the magma is created in the slab, between the slab and overlying mantle, or in the upper mantle beneath the crust. In addition, it is difficult to understand the mechanism that moves the fluid generated toward the Earth's surface. Is it hot enough to cause melting of rocks as it migrates toward the surface, thus changing its chemical composition, or does erupted lava represent the original melted rock that formed at or near the subduction zone? Regardless of which set of conditions is correct, the result at the Earth's surface is the same – lava erupts to create volcanic mountain chains, often forming composite volcanoes, primarily of andesite, dacite, and/ or rhyolite. It is just such a string of composite cones that makes up the High Cascades we see today.

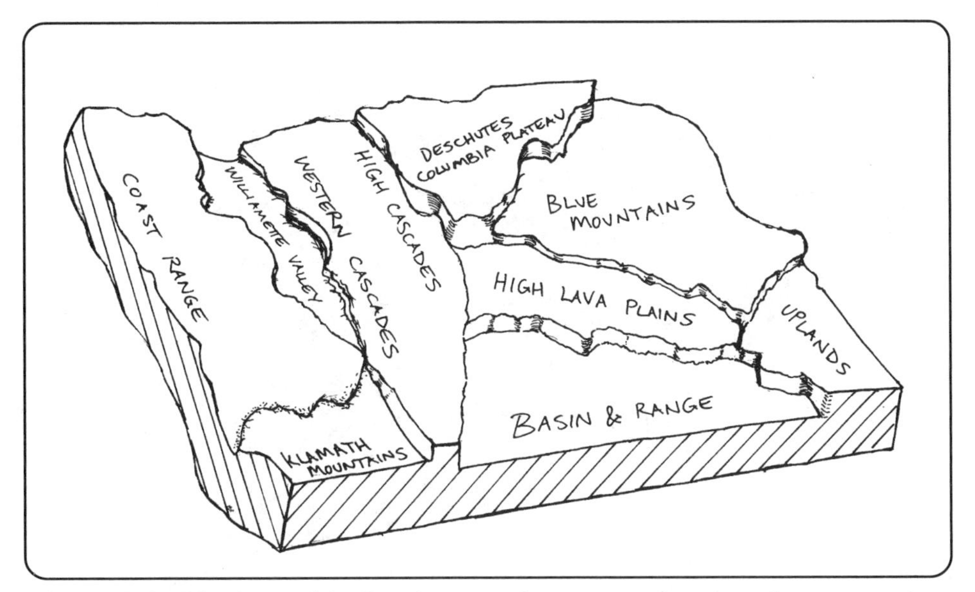

Figure 3-4 Physiographic Provinces *– Oregon can be viewed as ten regions that represent the distinct geology of each area. (After Orr and Orr, 1999.)*

Oregon Geology

Although there is much geological history that can be drawn into a discussion of Crater Lake, we will confine our discussion to a general summary of Oregon's geological history and the background necessary to understand why and how the High Cascade Volcanic Range formed. Very few, if any, of these details can be worked out by looking at the rocks in Crater Lake National Park. Rather, the story has been pieced together by many investigators examining geological features across the region and even on the adjacent ocean floor. With this diversity of evidence and the many investigators involved, there is a good deal of uncertainty about the complex geology seen across Oregon and adjoining states. This does not prevent geologists and others from attempting explanations, but most of these are best viewed as proposed models to see how well they fit the real geology. So, when reading, or hearing any geological explanation (including this book), try to keep in mind the tentative nature of such matters.

From a geological perspective the state of Oregon can be viewed as a collection of ten physiographic provinces, as illustrated in Figure 3-4. Each of these has a separate geological history and associated age as the western edge of North America developed. With Crater Lake located in the High Cascade volcanic chain, it is part of one of the youngest regions in the state. The geological history of many of these provinces, but not all, is intimately associated with a subduction zone and its related volcanic activity.

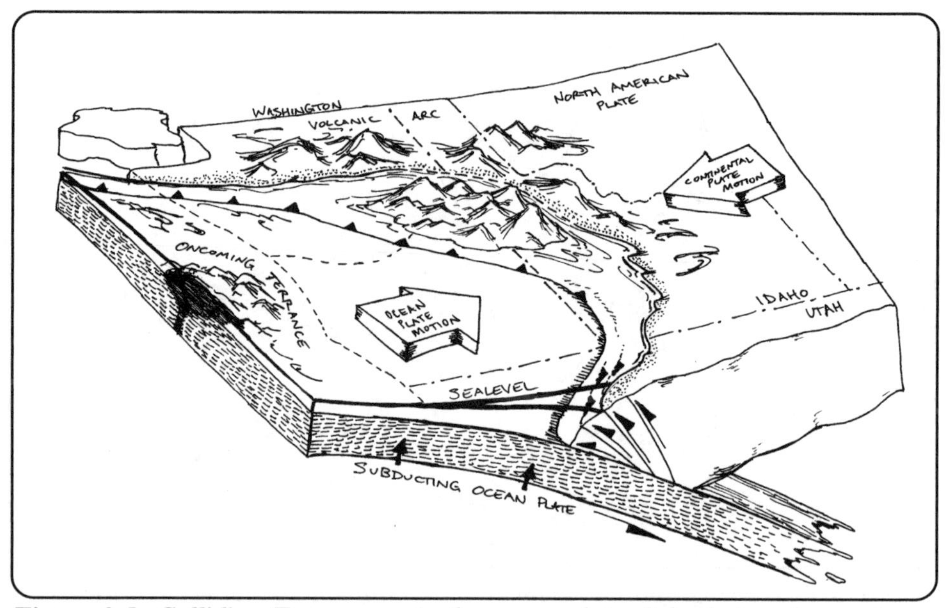

***Figure 3-5 Colliding Terranes** – As the ocean plate slides below the continent, a massive block is accreted to the North American plate. (After Orr and Orr, 1999.)*

Another important concept related to colliding continents and their subduction zones is accreted terranes (Box 3-2). Terranes, a recently developed concept, are crustal blocks bounded by faults with a geological history distinct from the surrounding areas. Field geologists have identified many of these terranes, especially in the state's earlier geological history.

Box 3-2
Exotic Terranes

A terrane is an area composed of related rocks that formed together but are much different from those surrounding them. They range in size from a few tens to a few thousand square kilometers (square miles). They may have been an oceanic island, part of the ocean floor, or even a piece of another continent. Terranes tend to be separated from nearby rocks by faults and have a distinct geological history including their age, paleomagnetic characteristics, rock type, and fossils they contained. The interpretation is that terranes have formed at another location and were carried along and attached to a continent (accreted) by a moving portion of an ocean plate; thus they are exotic to the location where found.

While most of the oldest rocks in Oregon date back to early Mesozoic time (starting about 200 Ma ago), there are some of Devonian age (400 Ma old). These rode into Oregon as terranes from areas in the ancestral Pacific Ocean, perhaps formed originally as volcanic islands. During this time the state lay under an embayment of the ocean

***Figure 3-6 Columbia River Gorge** – These fissure lava flows are up to 1.5 km (1 mi) thick and buried 165,000 km^2 (~62,000 mi^2) of the Earth's surface. Layer upon layer of lava can be seen in the gorge, but these never extended as far south as Crater Lake.*

with a shoreline laying across Washington, western Idaho, and northwestern Nevada. This ocean margin developed adjacent to a subducton zone, with the eastward-moving ocean floor diving below the continental mass. Thus, any terranes on the ocean plate eventually collide with the continental margin to add their mass to the North American plate (Figure 3-5). These rocks, formed from portions of small accreted terranes, have been identified by their boundary faults and dated by the fossils they contain.

With the start of the Mesozoic Era, the west coast continued to grow as more of these exotic blocks were joined to the continental margin. Both the Blue and the Klamath Mountains were added – much larger terranes than those previously accreted. As Cretaceous time arrived (80 Ma ago), two-thirds of Oregon was still beneath the ocean, with a subduction zone running northwest to southeast across its northeast corner. This began changing as the Cenozoic Era dawned (66 Ma ago): the shoreline had migrated westward and was then located from the Klamath Mountains through central Oregon and northward toward Washington State. At this point, a series of complicated events began that affected many parts of Oregon.

There were no mountains along Oregon's Pacific coast as the Cenozoic Era opened. A chain of volcanoes developed above the continental margin side of the subduction zone then lying west of the North American plate. This created an elongated offshore trench that accumulated volcanic material and sediments during the Eocene and Oligocene Epochs. As the ocean floor continued sliding beneath the North American plate, a huge variety of sediments accumulated in the subduction zone trench that was scraped off and added to the continent. Finally, starting at the southern end and progressing north, these rocks were compressed and uplifted to create the Coast Range. As soon as this began, in Miocene time, erosion started sculpturing the coast range – a process that continues even today. This major physiographic feature has a long and complex geological history with rocks composed of both volcanic materials and sediments.

Stepping back a few million years now, as the ocean shoreline and related subduction zone continued retreating westward, much of northern and central Oregon experienced two major episodes of volcanism. That portion of the state between the old and new subduction zones formed a stable platform extending the North American continent westward. This permitted other geological features to develop in what is now known as the Columbia Plateau Province. Starting about 36 Ma ago, a series of volcanoes erupted huge volumes of pyroclastics, ash, and dust, to form the John Day and other formations across much of Oregon. These were often in the form of ignimbrites and incandescent pyroclastic flows of gas-charged ash that traveled for miles across the land. Then, between 17 and 12 Ma ago (Miocene Epoch), the eroded surface of these deposits was buried under lava flows that filled valleys and low areas to form the Columbia Plateau (Figure 3-6). This feature, the second largest outpouring of flood basalts in the world, developed from numerous fissures that allowed magma to reach the surface as deposits of many thin basaltic flows. These eventually accumulated to hundreds or thousands of feet thick across 165,000 km^2 (~62,000 mi^2) of the Pacific Northwest.

Moving up to about 10 Ma ago, volcanic eruptions of lava and pyroclastics began in the High Lava Plains. All kinds of interesting features were produced: shield cones, cinder cones, tuff rings, lava tubes, and much more. These are related to a series of faults running in a NW-SE direction caused by two crustal blocks slipping past one another. Immediately south of the High Lava Plains, the Earth's crust is being stretched to create a series of north-south trending faults. Alternate blocks between adjacent faults were

Box 3-3
Faults, Grabens, and Horsts

When the Earth's surface is stressed, either pushed together, pulled apart, or twisted, rocks may fracture to produce a fault. In the High Cascades the stress is tensional (pulling apart), resulting in one side of a fracture slipping down relative to the other. If two adjacent breaks are facing one another, the region between them will be lowered to produce a feature known as a graben. The uplifted sections between two grabens is called a horst, a common feature in the Basin and Range Province southeast of Crater Lake. The Klamath Basin is a graben between two N-S trending faults.

Figure 3-7 Klamath Graben *– This view of the Klamath Basin looking south from Garfield Peak along the Caldera's south rim illustrates one of the grabens underlaying the High Cascades. The white peak in the upper right is Mount Shasta in northern California, about 150 km (100 mi) from Crater Lake.*

dropped down to form valleys (called grabens), while those remaining higher appear as mountains (horsts) today (Figure 3-7 and Box 3-3). The Klamath Basin, directly south of Crater Lake, represents one of these blocks that dropped along the northwest margin of the Basin and Range Province.

Cascade Province

Let's return back in time again to examine the Cascade Province. While the origin of the Cascade Range is not well understood, a number of models purport to explain its geological history. One of these suggests that some 40 Ma ago (Eocene Epoch), eruptions began from a chain of volcanoes lying along the ocean shoreline and ushered in this volcanic province.These volcanoes developed above an active subduction zone. Thick accumulations of lava and pyroclastic material were produced as the two crustal plates collided. Tilting and folding of these rocks in the Miocene was followed by eruption of more volcanics, resulting in continued growth of the Western Cascade chain. There is evidence of violent volcanism associated with the construction of composite volcanoes during this time. As activity in the Western Cascades during Pliocene time subsided, they were uplifted, folded, and tilted to produce an elevated region along its eastern margin. This increase in elevation allowed erosion by streams so that today these older Cascades appear as deeply incised ridges rather than the volcanoes that originally formed the range. This western portion of the Cascade Province has a long his-

***Figure 3-8 Fault Scarp** – The flat surface on the left is just north of Klamath Falls along the east side of highway U.S. 97. It is an exposed fault plane shown close up on the right. This low area is occupied by Upper Klamath Lake and sits in the down-dropped portion between two adjacent faults – a graben.*

tory, the oldest rocks are about 42 Ma, while the youngest are only 10 Ma. In addition, a comparison of volcanic activity between the two sections indicates that roughly six times as much volcanism occurred along the Western Cascades as in the much younger eastern region.

High Cascades

Finally, the stage was set for the rise of the High Cascades (Figure 1-8). Like the Western Cascades, composition of the first volcanoes that lie along the High Cascades were basaltic. In fact, estimates suggest that as much as 85% of this young volcanic range is composed of basalt. Prior to the volcanism that produced the High Cascades, a series of faults developed along the eastern margin of the Western Cascades. The area between related pairs of faults, some 20 to 30 miles apart, subsided to form grabens. Then, about 4 Ma ago, many overlapping basaltic shield cones were built in these grabens along this Eastern Cascade Province margin (Figure 3-1). They formed a foundational platform for the spectacular and much taller composite cones that now represent the High Cascade Volcanic Range.

As basalt was ejected at the surface in the grabens, further settling occurred. Apparently magma carried to the surface creates space below allowing more downward pressure and sinking due to the load of additional surface cones. Thus,

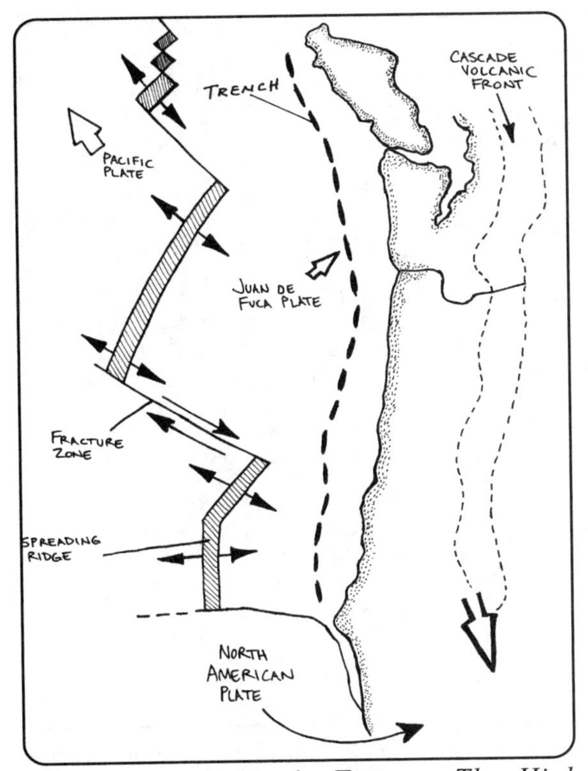

***Figure 3-9 Volcanic Front** – The High Cascades form a volcanic front resulting from the Juan de Fuca plate colliding with the North American plate to produce an offshore subduction zone and trench.*

tall composite volcanoes of the High Cascades were built in a series of depressions. The view south toward Upper Klamath Lake from the rim of Crater Lake illustrates part of the graben system. Much of highway U.S. 97 between Modoc Point and Klamath Falls follows the eastern marginal fault, with Upper Klamath Lake lying in the down-dropped portion of the graben. At certain locations along this route, well-exposed fault scarps can be seen (Figure 3-8).

These high volcanic peaks are Quaternary or younger and are primarily composed of andesite (Table 2-3). Dacite and rhyolite, the silica-rich volcanic materials, are found locally and are responsible for the explosive behavior of some High Cascade volcanoes. As a generalization, silica-rich eruptions are common toward the end of activity for the High Cascade composite cones. This was certainly the case for Mount Mazama during the climactic period that created the Crater Lake caldera.

Like much of the preceding geological history of the Pacific Northwest, the High Cascades are thought to form above two colliding crustal plates. An ocean plate, called the Juan de Fuca, is sliding beneath the western edge of the North American plate (Figure 3-9). Where temperatures and pressures produce melting, the magma rises to the surface, creating volcanoes of the High Cascades (Figure 3-10). Although eruptions along these young volcanoes have been rare over the past century, at least seven are known to have been active in historic time. Recently, of course, Mount St. Helens reminded us of their nature as it erupted throughout much of the 1980s and continues to be active. In Oregon, Mount Hood and the South Sister have recently displayed life.

For many millions of years, of course, the Coast Range, Western Cascades, and High Cascades have experienced significant erosion. A climate cool down

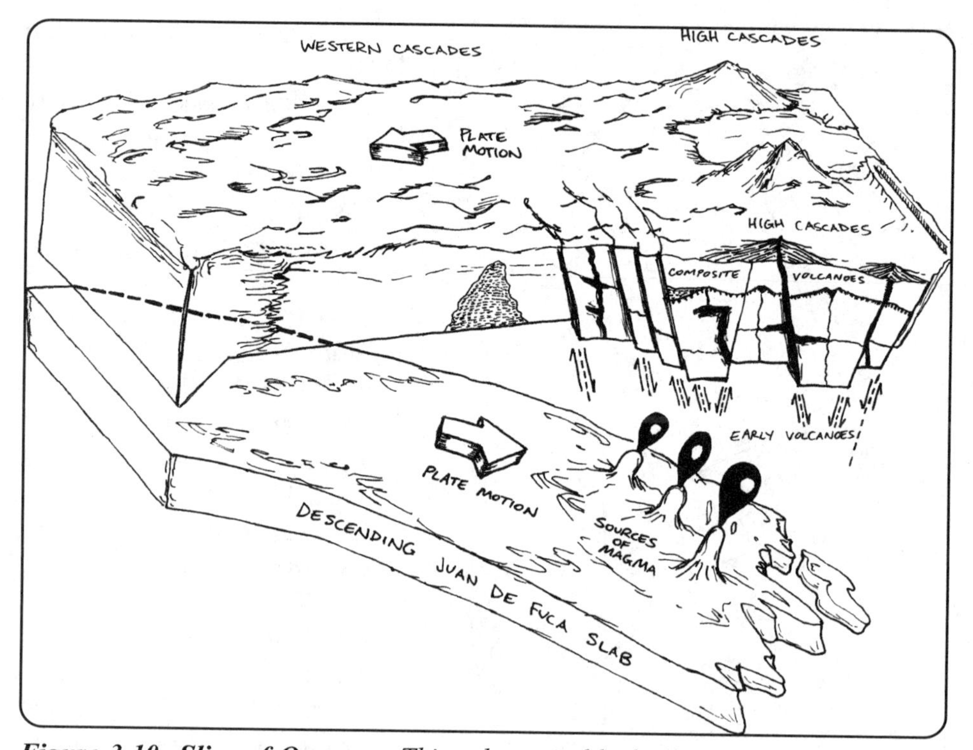

***Figure 3-10 Slice of Oregon** – This schematic block illustrates a cross section of the State from the High Cascades to the Willamette Valley. The graben structure below the High Cascade Volcanic Range and the volcano's source of magma originating near the downgoing slab is suggested. (After Orr and Orr, 1999.)*

ushered in the Quatenary Period about 2 Ma ago to produce glaciers over much of the Earth. Glaciers advanced into the region, especially affecting the higher elevations of the mountains. These extended over both the Western and High Cascades. As the large composite cones of the High Cascades were being constructed, glaciers were actively eroding them. At times, undoubtedly, products of both were being created and deposited together. Hot eruptive volcanic materials could cause melting of ice to produce mudflows (lahars) or other combinations of glacier and volcanic debris. These rivers of ice waxed and waned during most of the remaining time up to the present, resulting in ample evidence of their presence throughout the region. With warming in the last few thousand years, only a few glaciers remain in the Oregon portion of the Cascades – Mount Hood and the Three Sisters.

Additional Reading

Alt, D. D., and Hyndman, D. W., 1978, *Roadside Geology of Oregon*, Mountain Press Publishing Co., ISBN 0-87842-063-0, 268 pages. A handy refrence to Oregon's geology as viewed along the state's major highways.

Bishop, E. M., 2003, *In Search of Ancient Oregon*, Timber Press, ISBN 0-88192-590-X, 288 pages. A great collection of images and discussions that detail Oregon's geological history.

Harris, S. L., 1988, *Fire Mountains of the West*, Mountain Press, ISBN 0-87842-220-X, 379 pages. A good review of all the Cascade volcanoes, including a detailed chapter on Crater Lake.

Orr, E. L., and Orr, W. N., 1999, *Geology of Oregon*, Kendall/Hunt Publishing Co, ISBN 0-7872-6608-6, 254 pages. A detailed and technical description of Oregon's geology based on its physiographic provinces.

Chapter 4
Building Mount Mazama

Slightly east of the Western Cascades lies the most pronounced volcanic province in the Pacific Northwest – the High Cascade Range. Individual volcanoes along this chain have created some of the most spectacular scenery in the United States, if not the world. They were built on a foundation of older shield, composite, and monogenetic volcanoes along with associated valley-filling basalt to andesite lava flows (Box 4-1). Volcanoes north and south of Oregon tend to be higher than those in the middle portion of the chain. The two tallest peaks, Washington's Mount Rainier, at an elevation of 4,392 m (14,410 ft) and California's Mount Shasta, only slightly lower, at 4,346 m (14,261 ft), clearly illustrate this point. The high elevations created by this range are primarily responsible for the climatic variations observed in Oregon today. Moisture-laden air moving off the Pacific Ocean flows up and over the High Cascades, resulting in heavy precipitation on the west side, while desert conditions prevail to the east. To a great extent the landscapes we see today have developed under the combination of volcanic materials associated with this massive volcanic mountain system and the climatic extremes they have created. In this chapter we will investigate the evidence and processes responsible for the growth and modification of Mount Mazama, a volcano that formed one of the major peaks of the High Cascades.

High Cascade Volcanoes

Until recently, one of these large volcanoes, Mount Mazama, occupied the

Figure 4-1 Mount Rainier *– Over 25 glaciers presently move down the slopes of Mount Rainier, the highest of the Cascade volcanoes. Geological research indicates that past eruptions have created massive mudflows extending as far west as some of the most populated areas of Washington.*

present site of Crater Lake. This peak was constructed in much the same way as those still remaining. Learning how one was built, and the types of rock involved, will provide a reasonably good idea of the nature of any composite cone along the High Cascade Range. Crater Lake may offer a better opportunity to understand how these large volcanoes are constructed than any place in the Cascades.

When did the Cascade composite cones start forming? Just a few years ago, geologists would have answered, "about two million years ago." More recently, the number being used is 760,000 years or less. This rather dramatic change in dating the major Cascade peaks is not as unusual as it first appears. The older figure was based on estimating the age of volcanoes from geological studies in many different areas along the Range. By contrast, the 760,000-year upper limit is determined by measurement of the magnetic characteristics of entire lava sequences for a few Cascade volcanoes (Jefferson,

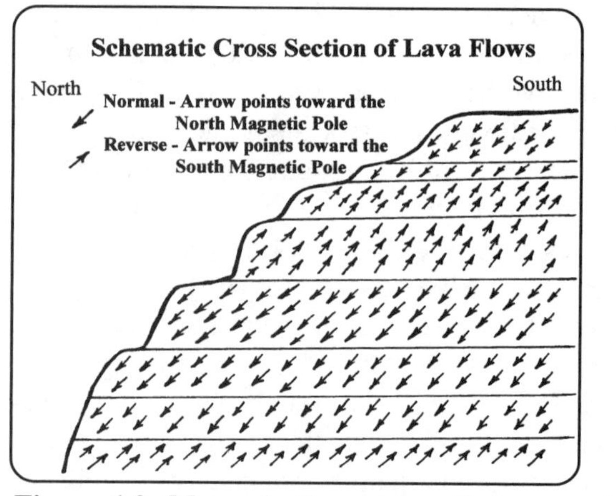

Figure 4-2 Magnetic Reversals *– Tiny magnetic crystals (-->) in lava flows record the orientation of the Earth's magnetic field. They align parallel to the field before the lava hardens to preserve their north-south directions. By examining the orientation of these natural magnets in this Northern Hemisphere view, the sequence of reversals can be determined.*

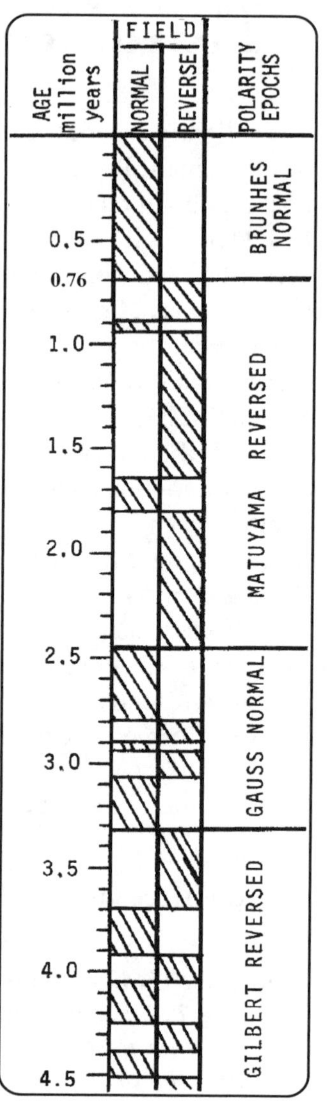

Figure 4-3 Geomagnetic Time Scale *– When the ages of magnetic reversals from a large number of lavas are combined, a distinct pattern develops. This scale is similar to the Standard Geological Time Scale (Table 3-1) but based on magnetic data.*

Three Sisters, and Mount Mazama). The technique used to estimate this younger age is based on the magnetic properties of volcanic rocks. Measurements based on tiny magnetic minerals, primarily magnetite, indicate that the Earth's magnetic field has reversed itself – switched directions – at certain times in the past (Figure 4-2). As lava cools to form solid rock, these small minerals act like miniature magnets that line up parallel with the Earth's magnetic field, just like a compass needle. When the Earth's magnetic field is normal, like today, the north-seeking end of these natural compasses point north. When reversed, however (flipped to the opposite direction), the north end would point toward the Earth's south magnetic pole. The last time the magnetic field reversed was about 760,000 years ago (Figure 4-3). Since all lavas produced by the High Cascade volcanoes have normal orientation, they appear to have formed during the present magnetic orientation – within the last 760,000 years.

Box 4-1
Monogenetic Volcanoes

The massive Cascade volcanoes have relatively long histories – often with many different types of eruptions. The most common volcanoes along the Cascade Range, however, are not these beautiful composite cones. They are small features known as cinder cones (also called scoria cones). Large ones are a few hundred meters high and about a kilometer or more in diameter. They have steep sides, up to 30 degrees to the horizontal, resulting from the explosive eruption of primarily lapilli-sized ejecta. Bombs and blocks (Table 2-1) are also common. In addition, lava flows are also produced, usually issuing from basaltic portions of the cone. The interesting characteristic of cinder cones is their relatively short life span. Most are active for just a few months, although a few continue for several years. Once eruptions cease, the volcano is dead and seldom is active again. Thus, volcanic regions including the Cascade Range, are littered with these small features. Composition is usually basaltic, prompting the name monogenetic suggesting one eruption and one source of volcanic material. However, similar but much less common features are composed of silica-rich ejecta known as pumice cones. There are more than twenty cinder cones in the Crater Lake area.

Mount Mazama, the volcano built where Crater Lake is now located, shares this age with the other composite volcanoes that make up the Cascade Range. Likewise, it shares the mode of construction of those high peaks as well as other similar volcanoes around the world. It may be helpful to divide Mazama's history and the geology of Crater Lake National Park and the surrounding area into two parts: (1) volcanic eruptions and other geological processes that occurred before the climactic events, and (2) those that followed that dramatic period. Most of the rocks exposed in the caldera walls belong to the first of these, which covers a period of nearly 400,000 years (Table 4-1).

Mount Mazama

Most of the volcanic material that made up Mount Mazama was ejected from a central or main vent area. This does not mean that Mazama had only one crater, for as we will see later, there were likely several vents in the summit area. This is the main reason that composite volcanoes, or any volcano, tend to produce an inverted cone shape as they grow in size. Another factor is the relative amounts of lava and pyroclastics produced by the volcano. And, as noted in Chapter 2, composite cones are built by a mixture of fluid rock and solid debris (pyroclastics), more or less alternating during the time of construction. Mount Mazama was no exception to this model, although it appears that more lavas were produced than pyroclastic materials. As a result, it may have been somewhat less steep than some of the other Cascade cones, which produced a greater percentage of cone-steepening pyroclastic material.

So far our discussion of volcanic materials has been uncomplicated with the question of whether the ejecta were either lava or pyroclastic debris. The real situation is more complex, for several processes modify these products after they have been produced and deposited on the volcano's slopes. Although ejecta can

Figure 4-4 Mount St. Helens *– During the 1980 eruption, heat melted glacier ice, creating a massive mixture of rock, mud, and water that swept down the northern slopes of the volcano. This mixture, known as a lahar, flowed down the Toutle River for many kilometers (miles).*

Table 4-1
Chronological Development of Mount Mazama*

Date/Age	Feature/Location	Source**
27,000 BP	Redcloud Cliff and Grouse Hill	1, 2, 3
36,000 BP	Fragmental deposits at top of Corkscrew (dacite)	1, 3
50,000 BP	Watchman Flow (dike?), andesite outcrops along Rim Drive west of Pumice Point, glaciated andesite lava flow at base of Pumice Point	1
> 70,000 BP	Lavas in SW caldera wall older than Hillman lavas	1
~70,000	Pumice Castle dacite (dates on lava flows above and below)	1
100,000 BP	Andesite lava flow below glacial till on Cleetwood Trail	1
110,000 BP	Palisades flow (dacite)	1
170,000 BP	Lava at lake level below Llao Rock	1
~220,000 BP	Garfield Peak – top two flows of andesite	1
230,000 – 340,000 BP	Andesites below dacites under Pumice Castle	1
~300,000 BP	Sentinel Rock dacite flows	1
~340,000 BP	Andesite below Sentinel Rock dacites	1
~400,000 BP	Dutton Cliff – rocks at water level and Phantom Cone (oldest on caldera wall)	1
~420,000 BP	Mount Scott (oldest dated lavas of Mount Mazama)	1
420,000 – 500,000? BP	Pre-Mazama rhyodacite domes (Bear Butte, Lookout Butte, Scout Hill)	1

* summarized from C. R. Bacon, 1987, and other sources by Bacon.
** 1: K-Ar data, 2: Radiocarbon, 3: Field relations and paleomagnetic measurements.
BP = years before present. See Boxes 4-2 & 4-3 for background on how dates are determined.

be reworked or redeposited in many possible ways, we'll discuss only the major types. These include glaciers, running water, and mass wastage, the downhill movement of loose material due to gravity.

Glaciation

During the time that Mount Mazama and the other High Cascade volcanoes were being constructed, they were also being eroded by glaciers. During the last 200,000 – 300,000 years, there have been four major glacial advances in North America. At higher elevations, such as the High Cascades, this meant that huge amounts of ice accumulated and moved downhill. At times, eruptions would produce volcanic ejecta that deposited material directly on the glaciers rather than on the rocks below. If these eruptions consisted of hot fluid lavas, portions of the ice would melt and, quite likely, result in explosions as melt water was rapidly converted to steam. Large amounts of water produced during such events often resulted in catastrophic events: massive flood-like runoffs or mudslides. A minor version of a similar process resulted in the devastated area at California's Lassen Peak during the 1917 eruption. Of course, the May 1980 eruption of Mount St. Helens provided a modern example of this phenomenon (Figure 4-4). Similar or larger events surely occurred as Mount Mazama was being constructed, but evi-

***Figure 4-5 Munson Valley** – This view down on Park Headquarters and the Steel Center from Garfield Peak illustrates the U-shaped, flat-bottomed nature of this glacial valley. The entire Rim Village area was beneath ice when the Munson Glacier filled this valley.*

***Figure 4-6 Glacial Striations** – In the past, Mount Mazama had glaciers flowing down its flanks that caused scratches and polish. Small patches still remain in several locations along the rim. This example can be seen near the North Junction on Rim Drive. (Note the knife for scale.)*

dence of such events is difficult to identify now. Much of the pyroclastic ejecta originally deposited on any existing glaciers was carried away from the central vent, only to be left far down the valleys.

Much of the Crater Lake landscape has resulted from glacial erosion as well as volcanic activity. The Pleistocene Epoch, noted in Table 3-1, is commonly called the Ice Age, for during this time glaciers were widespread in the Northwest. Indeed, the Cascade volcanoes were periodically buried in moving tongues of ice. When the climate was cool enough to allow winter accumulations of snow to build up from year to year, glaciers would form and grow. Temperatures and snowfall today still nourish remnants of these icy rivers, especially in the Washington section of the Cascades (Figure 4-1). Farther south, however, few glaciers exist, and their presence in the past must be inferred by other evidence.

At Crater Lake, direct evidence for valley glaciers may be found for much older episodes, as well as those responsible for shaping the present surface. One of the primary indications of glacial activity in mountains is the U-shaped valleys that display broad, flat bottoms. The cross section of Kerr Notch and Sun Notch as viewed across the lake are near-perfect examples, as is Munson Valley looking down from the Garfield Peak Trail (Figure 4-5). Other low sections of the caldera rim may also have been locations where glaciers developed and flowed away from Mazama's summit area.

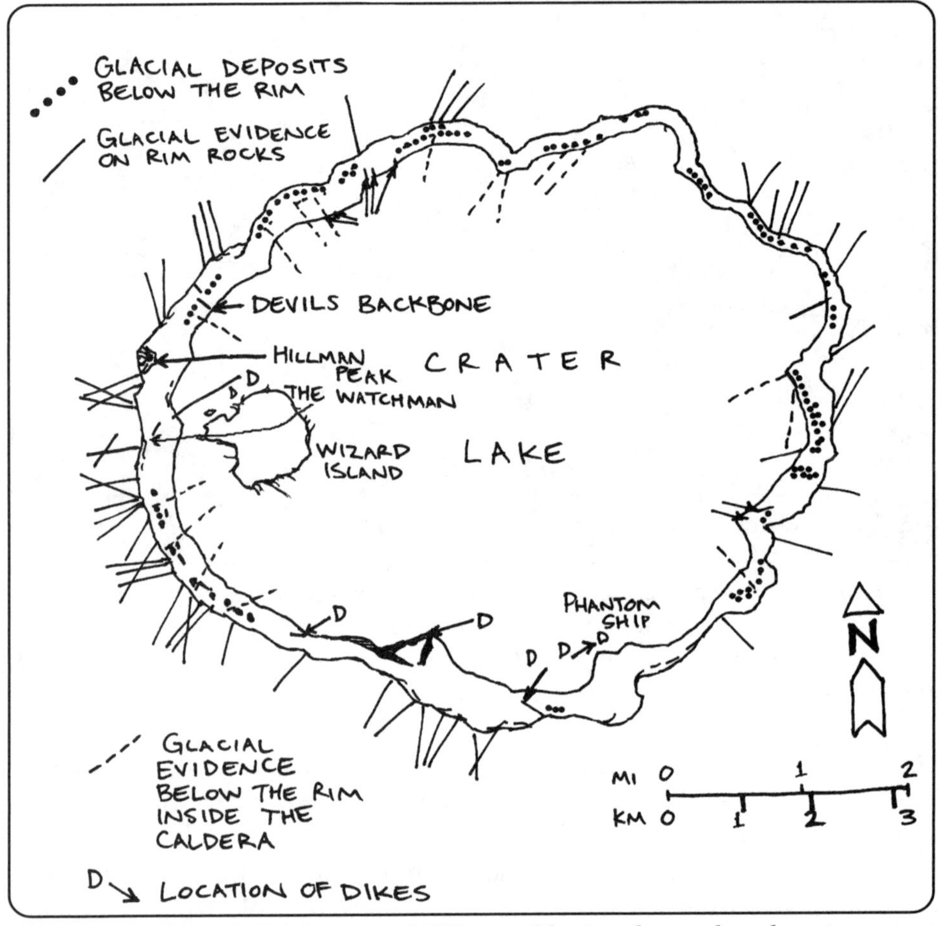

Figure 4-7 Glacial Evidence and Dikes *– Glaciers have played an important role during Mount Mazama's growth. Some of the glacial evidence presented on this map, however, has been reinterpreted and is now thought to represent non-glacial deposits. (After Atwood, 1936, and Williams, 1942.)*

Direct evidence of glacial activity is also indicated by the glacial striations (scratches) found on the rocks exposed along the caldera rim (Figure 4-6). These are caused by rocks in the bottom of the ice being dragged across the surface in a sandpaper-like fashion. If the rock material being carried is fine, a smoothing or polish-like finish may develop, rather than coarse notches or gouges. Regardless of the size of these features, they provide positive evidence of glaciers and the direction ice was flowing as it moved downhill (Figure 4-7). Although volcanic rocks may not accept or retain glacial polish as well as other rock types, this effect can be clearly seen in several places. Rocks exposed near Discovery Point and at North Junction provide good examples with easy access.

***Figure 4-8 Llao Rock** – More than half the height of the caldera wall here consists of a single lava flow. This was originally thought to fill a large glacial valley, but recent field evidence suggests the depression is an explosion crater.*

Many of these glacial indicators can be seen inside the caldera, presenting evidence for glacial activity as Mount Mazama was being constructed. But any surface developed while Mazama was actively growing, such as the U-shaped valleys, will be buried beneath materials deposited from later eruptions or left by retreating ice. There are examples of buried glacial valleys exposed in the walls of the caldera. The procedure for establishing the presence of a glacial valley requires more than merely finding an appropriately shaped cross section. Positive evidence, like polish, striations, or glacially deposited material, must be found to confirm the presence of glacial activity. Geologists in the past may have been too eager to assign some features to a glacial origin. An obvious example is Llao Rock (Figure 4-8), which does have a U-shaped appearance. Recent studies suggest a completely different explanation for the shape of Llao Rock's "valley." Farther east, the base of Round Top is also thought to result from glacial activity. It probably flowed between ice tongues and appears to be underlain by glacial deposits. There is also evidence of glacial activity in some other locations along the caldera wall. The depression near lake level below the Palisades along the north wall and sections on the east wall are good examples. Detailed mapping inside the caldera indicates that glaciers were instrumental in Mount Mazama's construction.

Evidence suggests that recent glaciers were large and important erosional agents on certain parts of Mazama. Ice may have exceeded 150 m (500 ft) in places. Sun Notch and Kerr Notch testify to this. As glaciers flow, they transport loose rock material, ultimately depositing their load at the glacier's terminus, somewhat like a giant conveyer belt. The resulting features are known as moraines and serve as direct evidence of glacial activity. These mound-shaped features consist

Figure 4-9 Avalanche Deposits *– During periods of eruptive activity, avalanche materials in the form of hot debris were emplaced on Mazama's slopes and in valleys. Originally, these were mapped as glacial deposits. This roadcut, through an avalanche debris along the road across from Park Headquarters, was thought to be a small glacial moraine.*

of a mixture of rock sizes from extremely fine powder (rock flour) to huge blocks many meters (feet) across. Until recently, many of the exposures in the valleys radiating out from Crater Lake were considered to be glacial deposits. These deposits do have many of the characteristics found in moraines. Careful study by Bacon, however, has shown some of them to be composed of volcanic material, rather than glacial till (Figure 4-9).These "nonmoraines" were produced by explosive eruptions near the time of the climactic events and will be considered in detail in Chapter 5.

Water and Gravity

Due to its abundance, water is usually the most important erosional agent in a humid climate. This is undoubtedly true for Mount Mazama as it erupted, grew, and was altered by flowing water. A glacial valley starts as a stream valley that is certainly occupied by running water when not filled with ice. But evidence of erosion by water as Mount Mazama was being built is almost nonexistent at the surface today. With the advent of the climactic eruptions, all existing valleys were buried under many meters (feet) of pumice and other debris that completely concealed the original topography of the volcano. What is seen in the park now is water erosion of these young deposits, which has occurred in the last few thou-

***Figure 4-10 Avalanche Chute** – This open strip resulted from snow and rock debris sliding down the slope, a form of mass wastage. The large block rolled down from its original position along the caldera rim near Garfield Peak. This view is along the road a quarter mile above Park Headquarters.*

sand years. The formation of these spectacular gorges and how they relate to Mazama's history will be discussed in Chapter 6.

One of the most widespread agents of erosion is the downward pull of gravity (mass wastage). Since it is present everywhere on the Earth's surface, gravity only needs loose rock and a slope to be an effective transporting agent. Much of the volcanic material produced by a composite volcano is loose debris, so mass wastage is prevalent on these structures. Whenever the slope becomes too steep to maintain a stable condition, a portion will simply break away and slide downhill as a rockslide or landslide (Figure 4-10). Angles as low as 30 degrees, or even less, are steep enough to allow mass wastage to occur.

Factors affecting how and when mass wastage occurs include many conditions such as the type of debris and slope angle. Angular blocks tend to be less likely to move than more rounded shapes. Water, however, may be the most critical influence in allowing movement to occur. Generally, the greater the percentage of water in loose rock material, the easier it will respond to gravitational forces. As a result, water-saturated debris may move down slopes of only a few degrees as a semi-fluid mass. And, like the other evidence for surface features present prior to the climactic eruptions, these are generally buried under many meters (feet) of ash, pumice, and other volcanic debris. It should also be noted that rockslides and landslides may be triggered by earthquakes, a common event in volcanic areas – more on this matter when the future of Crater Lake is considered in Chapter 9.

Inside the Caldera

One of the reasons Crater Lake is so pleasing to the eye is its high walls, which confine the lake on all sides. This same feature provides much of the evidence for geologists who are trying to reconstruct how Mount Mazama was originally constructed. The rocks displayed in the caldera walls indicate a much more complex geological history for Mazama than the basic nature of a composite volcano suggests, as described in Chapter 2. Much of this evidence can be appreciated only from lake level. The boat tours offer one of the best methods for visitors to examine the geology exposed in the caldera walls.

To find the oldest rocks in Crater Lake National Park, however, we must look outside the caldera. Some of these are located a few kilometers (miles) to the east. Eruptions from at least twenty-five vents (for example, Pothole Butte and Lookout Butte) have produced silicic domes and flows dated at between 420,000 and 500,000 years. Closer in, rocks composing Mount Scott have been determined to be about 420,000 years old, the oldest of those composing Mazama's overlapping cones. Along the southeastern caldera wall, adjacent to the Phantom Ship, exposed rocks also fit into the category of some of the oldest in the park. Using modern dating methods (Box 4-2), a tentative age of about 400,000 years has been determined for rocks found near lake level in this area. These rocks probably do not represent the original eruptions of Mount Mazama but are part of an early volcano that occupied this site.

This small cone has been given the name Phantom Cone because the Phantom Ship (Figure 4-11) composes a portion of the feature. By observing the caldera wall below Dutton Cliff, a good cross section of this early structure can be seen, comprising about half the distance from water level to the rim. Much younger lavas lap upon the Phantom Cone

Box 4-2
Radiometric Dating Methods

A number of elements found in rocks are radioactive – these have unstable nuclei that decay to form another element. Although this process is random, the average time it takes for half of a given amount to change is well known, a length of time called its half-life. The key to using these methods is that the half-life for any radioactive element is constant. For example, the half-life for a commonly used element, potassium 40, is 1.5 billion years. By measuring the amount of various radioactive elements found in rocks, their age can be estimated. This allows Crater Lake rocks to be dated and placed in chronologic order (Table 4-1). One of the commonly used radioactive elements is carbon 14. While it has been used in dating the climactic eruption materials, it is limited and not useful for most of the volcanic rocks at Crater Lake.

Figure 4-11 Phantom Ship *– As the small Phantom Cone, which makes up the lower portion of Dutton Cliff, was intruded by magma, dikes were formed. The sails originally thought to be a dike of more resistant material have more recently been mapped as breccia (Plate 6).*

from the east, in a rather pronounced slope in this portion of the caldera wall.

Other parts of the caldera wall are younger than this southeastern region. The lower portions of the caldera are composed primarily of andesitic lava flows. Individual flows range from 3 to 25 m (10 to 80 ft) thick, but the majority are in the category of 6 – 10 m (20 – 30 ft). From the description of composite volcanoes in Chapter 2, alternate layers of lava and pyroclastic material would be expected in a cross section, and this is what a typical wall of the Crater Lake caldera looks like upon first examination. The massive dark zones appear to be lava, while the lighter, more tex-

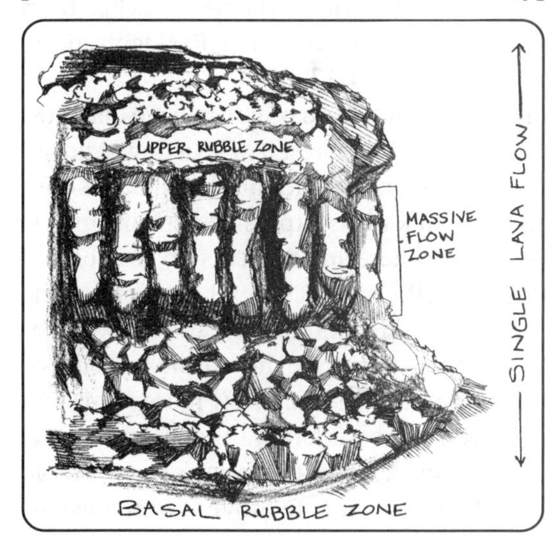

Figure 4-12 Auto-Brecciation *This cross section of a single lava flow shows the three zones that develop as the flow moves. Cooling at the base and the top causes the lava to solidify, then break up into rubble zones. The more fluid lava in the middle solidifies after all movement has ceased, resulting in the massive appearance, often forming columnar joints.*

***Figure 4-13 Lava Flow** – This flow along Rim Drive just west of Rim Village illustrates the middle and lower portion of an andesite lava flow.*

tured layers seem to be pyroclastic debris. In some places this is the correct interpretation, but there is another explanation for much of the alternate layers seen in the walls (Figure 4-12).

When a lava flow moves down slope, the bottom will cool more quickly as heat is rapidly lost to the ground, and the liquid rock starts to harden. With the continual movement of the more fluid lava in the central part of the flow, this solidified layer on the bottom tends to break into blocky pieces, a process known as auto-brecciation. As a result, when a cross section of one of these flows is observed, it may appear to be composed of two zones, a lower pyroclastic-looking region and a lava flow on top. To add a further complication, the same process may occur in the upper part of the flow, creating a third zone on top. This results in a rapid loss of heat to the air and creates a blocky pyroclastic-looking zone on top of the massive-appearing lava below. In general, rubbly flow tops are much thicker than those at a flow's base. Much of the rubble at the base is typically the flow-top rubble of the underlying flow. From a distance it's difficult to distinguish these rubble zones that form the top and bottom of a flow from actual pyroclastic deposits. Roadcuts along Rim Drive illustrate this process quite well (Figure 4-13).

With this concept of auto-brecciation in mind, it is easier to believe that most of the layering visible in the caldera walls is actually the result of lava flows and not pyroclastic deposits. This leads to the conclusion that Mount Mazama was constructed primarily of andesite lava flows with far less explosive deposits than is suggested for the origin of composite cones. There are, however, examples that appear to be pyroclastic debris interlayered with lavas at several locations inside the caldera. Both Eagle Craigs (below Garfield Peak, Plate 7)) and the wall between Sun Notch and Applegate Peak seem to illustrate the classic layering of composite cones, but the actual situation is more complicated (Figure 4-14).

Perhaps the most obvious features seen in the caldera wall are the numerous dikes (magma that solidifies after filling in cracks). During Mazama's construction, stresses caused vertical fractures that allowed magma to move toward the sur-

Figure 4-14 Eagle Craigs *– This portion of the caldera wall appears to represent an ideal structure for a composite volcano. However, the situation is more complicated ,and most of these layers are really an unusual type of lava flow.*

face. In many cases this fluid rock poured out on the surface as a lava flow, but sometimes it became trapped below ground. In any event, these fractures were filled with hot rock, which eventually cooled to create the dikes seen along the caldera wall today. The most famous of these, called the Devils Backbone, is easily seen from many points along Rim Drive, as well as from Sinnott Memorial Overlook below Rim Village. The most striking view of this feature, however, is from the boats, where the full effect of its massive height and width can be thoroughly appreciated (Figure 4-15). At the base, near the water, its width is over 15 m (50 ft), and this portion rises to nearly 160 m (500 ft). It's one of the youngest of the many dikes, for it cuts across all the older material exposed in that portion of the wall. Logic dictates those rocks were present prior to the emplacement of the Devils Backbone dike (Box 4-3). Altogether there are at least twenty dikes exposed in the wall, many occurring along the northwestern and eastern portions of the caldera (Figure 4-7).

An especially interesting section of the caldera wall can be seen at the west

Box 4-3
Relative Dating Methods

While radiometric dating methods provide the age of rocks or geologic features in years, relative dating methods give ages in relation to one another. There are a number of relative techniques commonly used, but two are prominent in volcanic areas, based on deciding which rock or feature was formed first. **Superposition**, the youngest rock on top, is the most obvious. In a series of lavas, they become younger from bottom to top. **Cross-cutting relationships** are also important. Thus a feature, such as a dike, that cuts across another is younger. Another observation is also important when evidence indicates that erosion has removed rock material, resulting in an **unconformity**. This feature indicates some amount of time is not represented by volcanic material.

Figure 4-15 Devils Backbone *This largest dike exposed in the caldera wall is composed of two sections and extends from the lake level to the rim. It illustrates the pattern of radiating dikes associated with composite cones (Plate 11).*

Figure 4-16 The Watchman *– A dike leads up to the summit and represents the source of magma that fed The Watchman volcano vent (Plate 4).*

end of the lake under the two peaks named The Watchman and Hillman Peak. Figure 4-16 is a view of The Watchman from lake level. Even more interesting are rocks observed on either side of Hillman Peak where the horizontal layers are readily apparent, indicating a typical portion of the wall. Moving in below the peak, however, produces a dramatic change, for the regular layers give way to a dark, massive region with no pattern (Figure 4-17). This represents the internal structure of a volcano that may have been part of Mount Mazama's summit complex. Indeed, this evidence suggests that the summit region of Mount Mazama was composed of a series of smaller volcanoes like Hillman Peak. When the collapse occurred, approximately half of this volcano was caught in the portion that fell away, leaving a cross section of its vent region. Back from the caldera wall along Rim Drive, just north of The Watchman Turnout, a roadcut has been made through one of the lava flows that originated from the Hillman volcano. Looking at the caldera wall below The Watchman provides a similar situation. Rather than such a pronounced vent region, however, a dike can be observed that served as the source of fluid rock to feed The Watchman volcano.

The observant visitor may also notice brightly colored areas in several places inside the caldera. These usually occur as shades of yellow, buff, brown, or orange and often have a weathered appearance, more like an earthy texture than

***Figure 4-17 Hillman Peak** – Horizontal lava flows along the caldera wall are interrupted by jagged intrusions of magma filling the vent area of the Hillman volcano. The collapse dissected this cone and exposed its internal structure.*

volcanic rock. As eruptions were venting materials from the magma chamber below Mazama, hot, chemically active gases and solutions were also being released. As these moved upward through fractures, nearby porous rocks were attacked and partially decomposed, often changing the original minerals to clay-like materials. Minerals in these volcanic rocks containing iron were altered, resulting in color changes from dark green or black to the bright rust-colored material noted above. This activity is common in volcanic and hydrothermal regions wherever hot gases or water-rich fluids were present.

From the first extensive geological fieldwork at Crater Lake, a picture of what Mount Mazama looked like was based on other large Cascade volcanoes. Mount Rainier and Mount Shasta were used as models. This produced a picture of a more or less symmetrical cone with volcanic materials, lavas, and pyroclastics, ejected from a single vent (Figure 4-18). Based on these other peaks, Mazama was initially assumed to be on the order of 4,500 m (15,000 ft) high. With more recent field research, this number has been adjusted downward to about 3,600 m (12,000 ft). More important, however, is the nature of the summit area. Rather than a single vent so typical of Cascade volcanoes, Mazama's summit was likely a series of vents, perhaps with activity shifting from one to another from time to time. Hillman Peak and Mount Scott may represent the ends of a series of peaks stretched between them in an arc around the southern side of the summit area.

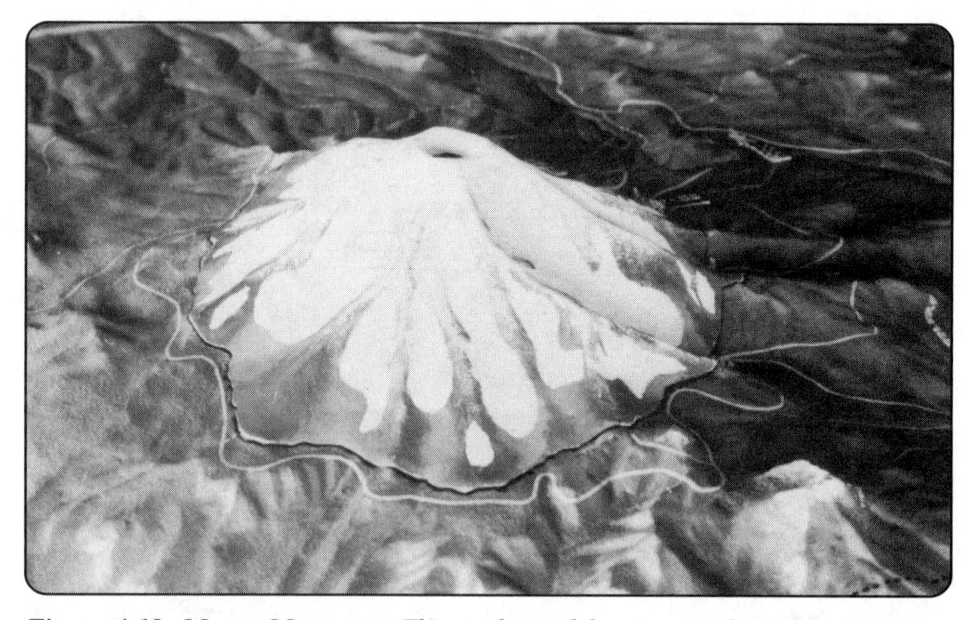

***Figure 4-18 Mount Mazama** – This scale model suggests what Mount Mazama may have looked like based on earlier research. Recent fieldwork suggests there were a number of vents making up the summit area. It also exaggerates the height of the summit. This view faces west on a model located in the Sinnott Memorial Overlook at Rim Village.*

Among the most striking features that can be viewed inside the caldera are the large talus slopes (Figure 4-19). These long, smooth, sandy-looking slopes are really secondary structures, resulting from mass wastage. They have developed from loose debris sliding down the caldera walls, most likely from young zones of massive pyroclastic accumulations. These talus slopes and other similar mass wastage features, commonly found in mountainous regions, are good illustrations of what happens when slopes become steep and unstable. From a distance they seem to offer easy access down to the lake but are much steeper and more difficult to climb than appearances suggest. By far the best, and only approved, route down to the lake is the Cleetwood Trail, which heads along the North Rim Drive. These talus slopes offer a glimpse of one of the events that occurred following Mazama's collapse; more about such features in Chapter 6.

Most of the evidence for Mazama's construction has been obtained from a careful study of the interior of the volcano (the caldera walls). This part of the park has recently contributed important new evidence to the next phase of this geological study. As noted in Chapter 1, early geologists hypothesized that Mount Mazama ultimately collapsed to form the basin that contains Crater Lake. How did this happen? More importantly, why? In the next chapter, the "real" story of Crater Lake will start to unfold as we investigate the final days of Mount Mazama.

Additional Reading

Bacon, C. R., 1987, Mount Mazama and Crater Lake caldera, Oregon, in *Geological Society of America Centennial Field Guide – Cordilleran Section,* v.1, p. 301-306. An 80 km (50 mi) road log describing the geology of Crater Lake National Park with detailed mileage and descriptions for seventeen stops – mostly along Rim Drive.

Harris, S. L., 1988, *Fire Mountains of the West – The Cascade and Mono Lake Volcanoes*, Mountain Press Publications Co.,ISBN 0-87842-220-X, 379 pages. An excellent reference covering the nature and history of the Cascade Volcanic Range.

***Figure 4-19 Talus Slope** – This view below Dutton Cliff above the southeastern shoreline of Crater Lake illustrates the steepness of the caldera walls. Weathering and erosion of loose pumice zones create these features in many locations around the basin.*

Chapter 5
Final Events at Mount Mazama

Mount Mazama's fate has been known for more than a hundred years, since Diller and Patton laid out the basic geological story in the 1890s. Crater Lake marks the location of a large Cascade composite volcano that erupted violently and ultimately collapsed upon itself to form a caldera. Perhaps the most eloquent description of this event was written by Howel Williams in his classic 1942 treatise on the geology of Crater Lake National Park. Since then, however, improved theoretical understanding of volcanoes coupled with additional field studies in the park have provided significant details of this amazing volcanic feature.

Williams' interpretation of the formation of the Crater Lake caldera is based on the large amounts of pumice found in and around the park. He believed that after a rather long rest period, probably lasting hundreds or thousands of years, Mazama's eruptions began again. Early activity in this renewed series of eruptions was not catastrophic. With time, however, the intensity increased as violent explosions produced small pumice fragments that could be transported away from the vent by the wind. As the size of these particles increased, they were not projected as high into the air, and the wind could not continue to serve as a transporting agent. Recognition of two distinct types of pumice, small and large particle deposits, provides evidence for two modes of emplacement. Smaller particles dropping out of the air represent early eruption stages, while the much larger sizes document

Figure 5-1 Granitoid Bombs *– These rare rocks are associated with the pyroclastic deposits found around Mazama. Older rocks lining the vent were caught up in the explosive eruptions and heated to high temperatures. This resulted in expansion that, upon cooling, caused cracks on the outer surface – a "bread-crust" effect (Plate 8). (Note the 15 cm [6 in.] ruler at lower right.)*

Figure 5-2 Williams Crater – *This cinder cone, within 1 km (0.6 mi) of the caldera rim, appears to have tapped materials from Mazama's magma cham-*

pumice flows formed later. Williams supposed that with this violent series of pumice ejections, the magma chamber was partially emptied, and the upper portion of Mount Mazama collapsed into the resulting void.

The insight Williams displayed is even more amazing in light of recent progress in understanding caldera formation. Much of the information described in this chapter and the next was not available to him as he pieced together the geological story at Crater Lake in the late 1930s. Today, additional geological field research, volcanic theory, and published information fills in many details. Much of this has been conducted by USGS geologist Charles Bacon over the past quarter century. He has greatly enriched the story of Mount Mazama's growth and final events at Crater Lake National Park.

Final Activities at Mount Mazama

Picking a time to begin this phase of Mount Mazama's geological history is arbitrary. A reasonable place to start is 40,000 years before the present, since that appears to be about the last time any major cone-building eruptions occurred. A period of rest followed which was interrupted when a series of eruptions ushered in the final climactic activity that resulted in Mazama's collapse and created the Crater Lake caldera. A number of interesting geological events occurred during this period (Table 5-1).

Table 5-1
Precursor Eruptions

Vent	Age	Composition	Volume	Length	Thickness	Air Fall	Emplacement
Cleetwood	7,700 yrs*	R H	~ 0.6 km³ (~ 0.15 mi³)	?	?	Yes	Explosive vent followed by a lava flow
Llao Rock		Y O	2.1 km³ (0.5 mi³)	1.5 km (1.0 mi)	Up to 375 m (1,200 ft)		Lava flow
	7,800–7,900 yrs*	D A	2-10 km³ (0.85 -2.4 mi³)	Wide spread	Up to 15 m (50 ft)	Yes	Explosive vent
Grouse Hill	27,000 yrs	C I	2.0 km³ (0.5 mi³)	~ 2 km (~ 1.3 mi)	~ 250 m (~ 850 ft)	No**	Lava flow
Redcloud Cliff	27,000 yrs*	T E	0.2 km³ (0.1 mi³)	1.5 km (1 mi)	Up to 200 m (~375 ft)	Yes	Explosive vent followed by a lava flow

* Calendar (cal) years before present (BP) based on radiocarbon dating.

** None identified, probably an artifact of exposure and preservation.

Here is the likely sequence. (1) High on the southwestern flanks of Mazama, a dome(s) developed and failed, producing debris flows over a wide area of the volcano, especially well exposed above Munson Valley. (2) A small cinder cone, Williams Crater (Figure 5-2), formed along the western margin of the caldera (formerly called Forgotten Crater as it had not been named before Williams' work). It appears to offer insight into the nature of Mount Mazama's internal plumbing. Some of the other cinder cones scattered throughout the park may also fit into this time period, although most appear to be older. (3) Most exciting, though, is the series of precursor eruptions that marked the beginning of the end of Mount Mazama. Eruptive centers at Redcloud Cliff, Grouse Hill, Llao Rock, and Cleetwood tell a fascinating account leading down to the time immediately before Mazama's cataclysmic climactic event (Figure 5-3). Williams included these precursor eruptions in his 'Northern Arc of Vents' (Williams, 1942, ps. 44-54) and outlined their history. In recent work, Bacon has greatly expanded our understanding of these features by adding important details.

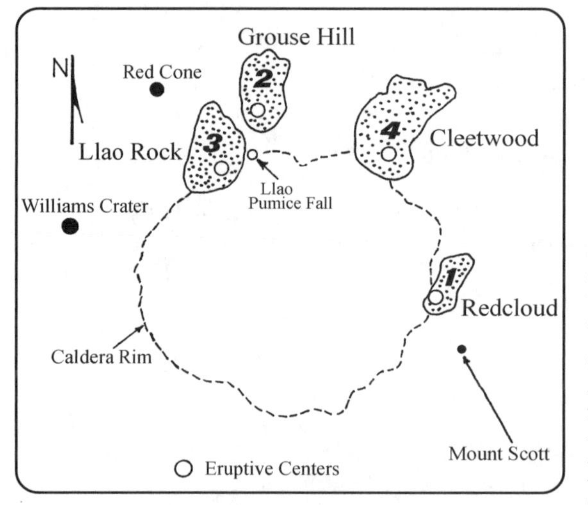

***Figure 5-3 Precursor Vents** – The location of the four eruptive centers preceding Mazama's climactic eruption. Numbers indicate the order in which they were active.* (*After Bacon, 1982.*)

Dome Emplacement and Debris Flows

A number of fragmental deposits are exposed in several locations on the southern slopes of Mazama. A relatively thick debris flow is overlain by thinner pyroclastic deposits. These have been interpreted as products from the destruction of a dome(s) by the injection of new lava (known as juvenile) during an explosive event(s). The debris-flow deposit represents the remains of the dome(s), while the overlying pyroclastic zones were emplaced as the new material moved downslope. The dome(s) appears to have formed at an elevation above 2,400 m (8,000 ft) high on the upper slopes or at the volcano's summit area.

This eruption(s) may represent one of the last events before the rest period preceding the climactic period. Based on their distribution, Bacon judged a common source area was located on Mazama's surface "east of Wizard Island and north of Garfield Peak." Using relative dating methods and paleomagnetic techniques (Box 5-1), they appear to be between 22,000 and 50,000 years old. While there are exposures of debris-flow deposits on Garfield Peak, near the Watchman and Devils Backbone (Plate 11), some of the best examples occur in Munson Valley.

Box 5-1
Paleomagnetic Methods

Certain rocks that contain iron minerals, like magnetite, become weakly magnetized in the Earth's magnetic field as they cool. These minerals behave like tiny magnets and their orientation and other properties can be measured. Their orientation, either normal or reversed (Figure 4-2), and other details provide information on magnetic conditions when they formed. By comparing these conditions with those determined for a certain period in the geological record, a tentative age may be suggested. Unlike radiometric methods, paleomagnetic measurements do not provide unique dates. Because similar magnetic conditions may occur at various times, paleomagnetic results are stated as "are consistent with" a certain time.

Apparently the dome(s) was disrupted. Instability of its steep slopes may have "uncorked" the core of the dome(s) or depressurized the feeding conduit. In either case, juvenile lava escaped, which sent debris cascading downslope in an avalanche-like fashion. The material was still hot and contained gases. This resulted in fumarolic alteration, causing bleaching and oxidation in some locations. Within these deposits are some of the most interesting rocks found in the park. Much of the lower debris flow is porphyritic dacite that composed the dome, but there are also blocks with andesite inclusions. The pyroclastic layer(s) on top have a more diverse composition, with blocks of dacite, like those beneath, along with rhyodacite and prismatically jointed blocks of dacite from the new lava (Figure 5-4).

Figure 5-4 Pyroclastic Deposits *Blocks of dacite from a collapsed dome were embedded in ash and deposited at high temperatures (Plate 17).*

In addition, the explosive materials contain pieces of preexisting rock caught up in the magma as it moved up through the volcanic pipe toward the surface. These are eventually coughed up with the other pyroclastic materials and are known as lithic fragments. They may be many different rock types, such as granite, which is not found in the rocks that compose Mount Mazama. In rare cases these lithic fragments have become so hot that they have expanded creating surface cracks, an appearance that looks much like breadcrust on volcanic bombs (Figure 5-1).

Box 5-2
Cascade Cinder Cones

Hundreds of cinder cones are scattered among the large composite volcanoes of the High Cascades. These monogenetic features typically have an extremely short active life. Most have an eruptive history of less than a year, thus, they represent an instant in geological time. While many appear symmetrical, others are elongate, reflecting the prevailing wind direction or trend of the fissure that provided their source of lava. Summit craters are common and tend to be large in comparison to the entire edifice. When fresh (young), they display crisp profiles. With age, however, cinder cones soften as expected due to erosion.

***Figure 5-5 Red Cone** – This feature along the North Entrance Road represents the many cinder cones found in and around Crater Lake National Park. These monogenetic volcanic cones were not related to Mount Mazama's magma chamber and appear to be of various ages.*

***Figure 5-6 Mixed Magma Bomb** – This bomb, erupted from Williams Crater, illustrates the mixing of basalt and dacite materials as Mazama's magma was tapped.*

Cinder Cones

Some twenty-plus cinder cones have been mapped in and around Crater Lake National Park. These are basalt and basaltic-andesite in composition and represent monogenetic features like hundreds of others that dot the Cascade landscape (Box 5-2). Magma sources for these small volcanic features are thought to derive from the Earth's mantle and are modified as they interact with crustal rocks during migration to the surface. Based on this interpretation, rocks in these cinder cones may represent compositions similar to those rocks that underlay Mount Mazama and other High Cascade volcanoes. Red Cone and Desert Cone are good examples of prominent cinder cones at Crater Lake (Figure 5-5). Wizard Island, located in the western end of the lake, is the most obvious cinder cone in the park. It is not related, however, to those outside the caldera as it is much younger and formed later from Mazama's magma chamber. More on Wizard Island in Chapter 7.

Williams used the term parasitic cones to describe these cinder cones. This suggests they are related to Mazama's magma chamber. Indeed, Williams (1942) clearly describes the nature of these small features when they occur "... on the flanks of a composite volcano (they) are usually a sign of old age." He further offers a lack of glacial effects on these cones as evidence for their relatively young age. Recent dating techniques, however, suggest these features have a range of ages, but further study is necessary to work out a detailed chronology. The relative ages along with other evidence, including glaciation of virtually all these vents, indicate most did not derive magma from the same source as Mount Mazama.

In one case, however, Williams Crater, near the west caldera wall along the Rim Drive, does appear to be related to the main magma chamber. This cinder cone complex, with a fresh appearance, consists of a basaltic lava flow and cinder cone formed prior to a mixed lava flow/dome eruption. These developed along a radial fissure, marking an inferred fracture system, extending to the northwest away from the caldera.

The critical evidence showing that Williams Crater is associated with Mazama's magma chamber is the nature of the final eruptive materials – mixed lavas. Apparently, the basaltic source for this cinder cone moved upward through a conduit near the margin of Mazama's magma chamber, which contained silica-enriched magma. Toward its final eruption stage, some of this higher silica magma 'leaked' along the radial fracture and mixed with the basalt to produce bands of dacite now found in basalt of the final flow. Most spectacular, however, are mixed bombs ejected during construction of the cinder cone phase (Figure 5-6).

Precursors to Mazama's Final Days

As Mazama entered this late stage of its activity, for approximately the past 40,000 years, the composition and nature of the magma occupying its chamber began to change. Throughout most of its history, the magma issued during eruptions was basaltic to andesitic. Although there were times when pyroclastic materials were produced, most of the volcanic activity was in the form of lava flows. The dacite dome(s)/debris flows noted above suggest this change – the magma composition was becoming silica enriched. This change in composition is clearly illustrated in the four vents that make up the precursor eruptions to the climactic event (Figure 5-3 and Table 5-1).

Redcloud Eruptive Center – The Redcloud center, now exposed along the northeast rim of the caldera, began with a pyroclastic eruption, leaving a pumice-

Figure 5-7 Redcloud Cliff – *This view from lake level illustrates the V-shaped nature of the explosion crater that initiated the precursor events.*

fall deposit. Evidence suggests a violent explosion produced the funnel-shaped crater now occupied by Redcloud Cliff (Figure 5-7). This was followed by the pumice fall that can be seen as a subtle lining below the massive Redcloud flow that fills the vent. It also forms a deposit along the road to Cloudcap (Figure 2-5). Once this gas-rich phase was exhausted, the viscous rhyodacite lava of Redcloud filled the crater and overflowed down the flanks of Mazama for about 1.5 km (1.0 mi). Banding in the flow suggests it was emplaced in a dome-like manner. In the vent portion visible on the caldera wall, the flow appears to be about 200 m (600 ft) thick, but to the east it thins to about 100 m (300 ft). Apparently the collapse that formed the caldera cut through the Redcloud flow just west of its vent where lava reached the surface. The Redcloud eruption materials appear to have a volume of at least 0.2 km^3 (0.05 mi^3), and adding in fall deposits would increase this amount.

Initially the craggy upper surface of the Redcloud flow exposed on the caldera wall suggested it had not been glaciated. This observation was used to estimate the flow's maximum age. Subsequent work, however, indicates the eruption occurred during the last glacial period.The radiocarbon age associated with Redcloud tephra eruption has been placed at 27,000 cal years BP.

Grouse Hill Lava Flow and Dome – The summit of this feature, located 2.75 km (1.5 mi) north of Llao Rock, is also rhyodacite. Apparently the initial lava flowed downslope about a mile from the vent. Lava also piled up to form a dome centered over the vent (Figure 5-8). Unlike Redcloud Cliff and Llao Rock, no pumice associated with the Grouse Hill eruption has been identified. Its surface is covered by a thick ejecta blanket of the climactic eruption material, and exposures are rare on top of the flow. However, steep cliffs can be seen along the margins. Diller considered Grouse Hill to be older than Llao Rock, but Williams thought they were about the same age. Recent evidence, however, places the Grouse Hill event at about the same time as Redcloud Cliff. The volume of all lava erupted at Grouse Hill is more than 2.0 km^3 (0.5 mi^3).

Mazama appears to have quieted down for the next few thousand years. The next volcanic activity is dated at about 8,000 years ago, when the first of two rhyodacite eruptions occurred from eruptive centers along the northern flanks of Mazama (Figure 5-3). These form the final precursors to the climactic eruption sequence that ended in the collapse of Mazama some 7,700 cal years BP.

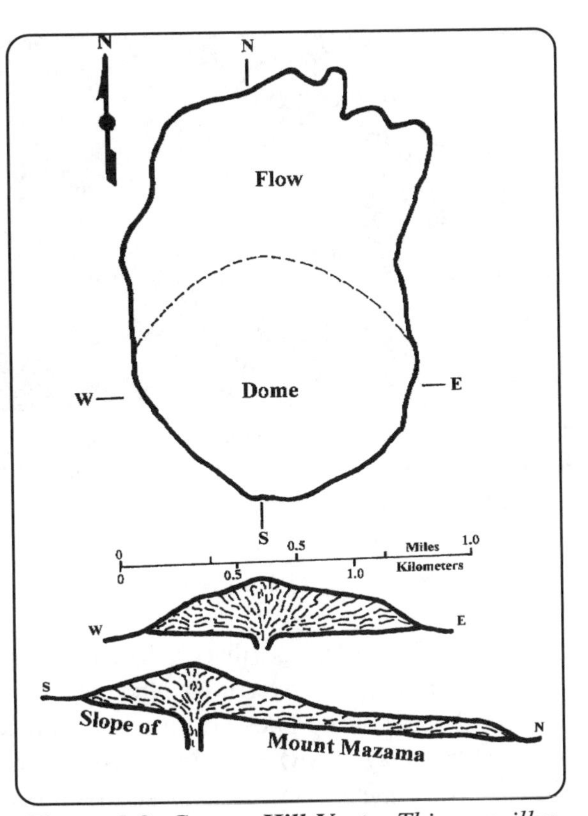

Figure 5-8 ***Grouse Hill Vent*** *– This map illustrates the internal structure and nature of the precursor vents leading up to Mazama's climactic eruption.* (*After Williams, 1942, Fig. 11.*)

Llao Rock Eruptive Center – One of the most distinctive features in the caldera wall, Llao Rock (Figure 4-8 and Plate 13), followed nearly twenty centuries after the Redcloud Cliff and Grouse Hill events. Based on glacial deposits found below the west wing of the flow, Williams thought the depression occupied by this dramatic flow was a glacial valley. Detailed work by Bacon, however, has reinterpreted this as an explosion crater formed during an initial phase of the eruption.

Llao Rock, like related preceding vents, has a rhyodacite composition and shows a similar eruption pattern as Redcloud Cliff. A large pyroclastic event spread a layer of pumice across the region, with ash beds found as far away as Washington, Nevada, and eastern Oregon. To distinguish this deposit from similar climactic pumice, it has been called the "Llao Rock Pumice Fall." In addition to its rhyodacite composition, this deposit is distinguished by mafic inclusions and up to 20% lithic

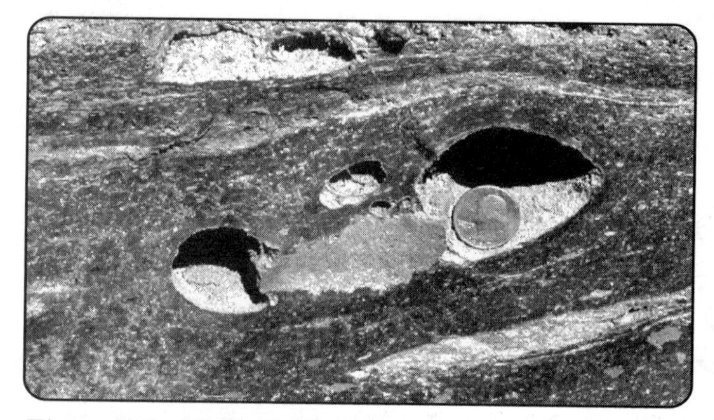

***Figure 5-9 Mafic Inclusions** – Iron and magnesium-rich inclusions are found in the precursor lavas to Mazama's climactic eruption, but are only common in the Llao Rock flow.*

fragments in some locations. These elements are similar to those in the overlying lava flow (Figure 5-9). A layer up to 15 m (50 ft) thick can be seen beneath the east wing of Llao Rock that contains coarse pumice blocks, some nearly 1 m (3 ft) in diameter. The upper portion of this deposit appears to be fused by heat of the overlying lava, showing columnar joints at its base (Figure 4-12 and Box 5-3).

With release of the gas that produced the driving force for the initial eruptions, viscous lava filled and overflowed the crater. Small dikes seen at the base of the flow, with the same composition as early-erupted Llao pumice, suggest a feeder source for the lava. However, closer inspection does not connect them with the flow. Paleomagnetic data also suggests they are older than the overlying lava flow. Thus, it's likely that the Llao Rock lavas reached the surface from a larger vent not visible in the face of the caldera wall.

The viscous rhyodacite lava of Llao Rock spread laterally some 1.8 km (1.25 mi) and moved downslope about 1.5 km (1 mi). In the center portion exposed in the caldera wall, it is some 375 m (1,200 ft) thick. The total volume of ejected lava is estimated to be 2.1 km^3 (0.5 mi^3), with possibly another 10 km^3 (2.4 mi^3) more in the initial pumice fall (Bacon estimates it between 2 and 10 km^3). A carbonized twig fragment found in a soil zone beneath the Llao Rock pumice fall yielded a date of about 7,800 to 7,900 cal years BP.

Box 5-3
Columnar Joints

Columnar joints, usually a vertical pattern of parallel cracks, are commonly seen in basaltic lavas but may occur in other volcanic materials. These features are a product of thermal stresses that cause contraction as fluid volcanic material cools to a solid. They tend to develop with five or six sides, but may be as few as three or as many as seven. The cracks form perpendicular to the cooling surface, usually near horizontal, and thus joints tend to be vertical. Columnar joints can be seen in many locations on the caldera walls and the climactic pyroclastic flow deposits in the walls of many of Mazama's eroded canyons.

***Figure 5-10 Cleetwood Vent** – This view of Cleetwood Cove illustrates Diller's "backflow." The wings of this flow are apparent on either side just below the caldera rim. Timber Crater, a large cider cone, is on the horizon.*

Paleomagnetic data obtained below the east wing of Llao Rock is consistent with this radiocarbon date. Climactic pumice completely blankets Llao Rock's upper surface.

Cleetwood Eruptive Center – This eruption occurred immediately prior to the climactic event at Crater Lake. In fact, the collapse of Mazama occurred so quickly afterwards, perhaps only months or days, that its lava had not completely cooled. Diller interpreted the jagged exposures of glassy rhyodacite in the caldera wall above Cleetwood Cove as lava flowing into the newly formed caldera – his so called backflow. Some later investigators did not accept this hypothesis, however, and suggested it represented the vent region of the Cleetwood flow – the lava was actually moving upward toward the surface. When collapse occurred, in this interpretation, it cut across the vent, exposing a cross section of the vent as part of the wall.

An explosive event ushered in the Cleetwood flow with a large pumice fall to deposit materials now exposed below the wings on either side of Cleetwood Cove. The trail to the boat landing passes through this material. Lava then issued from a vent a little northeast of Rim Drive (Figure 5-3). The composition of the rhyodacite lava is identical with the pumice produced during the climactic eruption, so it appears that both were drawing from the same portion of Mazama's magma chamber. There is additional evidence of how quickly the climactic eruption followed

***Figure 5-11 Welded Ash** – The Cleetwood flow was still hot (and fluid?) when the first air-fall of Mazama's Climactic eruption was deposited. The heat softened them and they stuck together (welded).*

the Cleetwood eruption.

Interpretation of the deposits in the roadcut through the Cleetwood lava is the key to understanding the timing between this flow and the climactic event. Most obvious is the dramatic red coloration caused by fumarolic gases originating in the lava. Next, careful observation reveals a thick layer of pumice overlying the lava, this is material produced during the climactic eruption. The upper part is loose, but the lower zone laying directly on Cleetwood's lava has been fused (welded) by the high temperatures of the escaping gas and lava. Thus, the Cleetwood flow must have been hot as the climactic eruption started, depositing pumice on its surface (Figure 5-11 and Plate 16). As it turns out, Diller got it right, and his "backflow" interpretation made over 100 years ago is correct.

Source of Mazama's Magma

In Chapter 2 various types of volcanic eruptive products and features were described. However, the ultimate supply of magma was not considered. How is magma formed, where does it come from, and how does it reach the Earth's surface? These are questions that have captured the attention of geologists, especially volcanologists, for several centuries – and still do!

Although definitive answers to such questions are sketchy, Earth scientists generally agree on some broad aspects about the source of magma. They assume the ultimate source of fluid rock is in the Earth's upper mantle and lower crust, some 30 km (20 mi) or more below the surface. The heat necessary to melt rock is thought to come from decay of radioactive elements along with the planet's original heat of formation. In the case of subduction zones, another possible source of heat is friction produced as the downgoing oceanic slab interacts with the underside of the overlying continental plate. Of course, pressure at these depths is also a factor, for higher temperatures are required to melt rocks at higher pressures. Based on the theoretical composition of the mantle, initial magmas are thought to be mafic, with high iron and magnesium and low silica content.

Once a body of magma has formed, it will migrate upward. This is likely a matter of density, for hot liquid rock is less dense than the rock it passes through. Just how this happens is not clear, although there are several possible mechanisms, including fracture zones, melting the rocks encountered, slender plumes of magma rising (called diapiric rise, Figure 3-10), or simply pushing near-surface rocks upward. In the High Cascades (and other volcanic fronts above subduction zones) the magma bodies tend to be localized for long periods, resulting in construction of large volcanoes. Moreover, the rocks that build these cones have a different composition – primarily andesite rather than the expected mafic magma (usually basalt) originally generated at depth.

Possible explanations for how intermediate andesite magma is produced are: (1) the original mafic magma has been altered by incorporation of silica-rich rocks near the surface, and (2) silica-rich sediment carried down with the subducting basaltic ocean plate is melted and mixed with the mafic materials along the subduction zone (Figure 3-3). Either way, andesites are erupted as lava flows at isolated locations to produce the large composite cones found along the High Cascades today.

A third possibility of producing intermediate and high-silica magmas is known as differentiation. As a mafic magma cools, crystals of certain minerals begin to form. These are rich in iron and magnesium and thus have a higher density than the remaining liquid. As they settle, the magma left is enriched in silica, and eventually the composition changes to andesite or even dacite or rhyolite. Differentiation is thought to be the main process that produced the pyroclastic materials erupted during the climactic event that created the Crater Lake caldera.

Regardless of how this happened, a supply of magma was produced and delivered beneath these volcanoes. However, it does not appear to be a uniform or continuous process, but rather episodic. Magma seems to arrive in "batches" that produce eruptions followed by quiet periods. This has led to development of the concept of a magma chamber, a shallow reservoir of magma that is replenished from time to time. This model is often used to explain the source of magma for a volcano's periodic eruptive behavior. And, of course, this concept reflects the geological nature of Mount Mazama (and the other High Cascade volcanoes) nicely – an eruptive period of andesite lava flows followed by a time of rest.

Since explosive volcanic activity is associated with subduction zones, Mount Mazama's construction on the margin of the North American continental plate overlying the Juan de Fuca downgoing ocean slab fits this model well (Figure 3-10). Composite volcanoes, developed above subduction zones, typically produce intermediate eruptive materials, primarily andesites. This, of course, is the history for Mount Mazama's eruptive life. Throughout most of its 400,000 years of growth, Mazama has produced andesite lava flows with an occasional explosive event yielding pyroclastic materials.

Assuming that Mazama had (and still has) a magma chamber that provided the materials for the climactic eruption, let's examine its nature. Apparently, its magmatic system formed at a location controlled by the structure of underlying rocks. It likely developed at a focus where magma, generated by the heating of deep crustal rocks, rose toward the surface. Batches of liquid andesitic rock (or slowly changing to andesite by differentiation) occupied a region below Mazama in a magma chamber and eventually erupted as lava. Occasionally, these eruptions would be explosive, as evidenced by pyroclastic zones found in the caldera wall. During these events the chamber contained magma with enough gas to power more violent activity.

This scenario of liquid rock residing in a volcano's chamber for a period of time, evolving into two or more distinct compositions, is an important concept. Their varying compositions are layered according to density, with the light, more silica-rich material accumulating on top of the silica-poor fluids. Another important factor is the relative high viscosity of high-silica magmas along with their tendency to contain more gases. The bottom line here is that they tend to produce explosive eruptions and may explain the pyroclastic materials formed during Mount Mazama's construction. More exciting, however, is how this relates to the climactic events that resulted in forming the Crater Lake caldera. That is the topic for Chapter 6.

Additional Reading

See Bacon (1982), Diller and Patton (1902), and Williams (1942) in the Appendix references.

Final Events at Mount Mazama

Chapter 6
Last Days at Mount Mazama

With the start of activity at the Cleetwood eruptive center, Mount Mazama's time was drawing to a close. The massive climactic event that would cause the mountain's collapse and the creation of the Crater Lake caldera was imminent. It was just a matter of days or weeks until Mazama and the surrounding landscape would be changed beyond recognition.

To understand what occurred during this short period of time, it is necessary to examine the nature of violent eruptions associated with composite volcanoes. And like many aspects of volcanic science, because real-time observations are not possible, much is based on theory and models. Of course, plinian columns (Box 6-1) have been observed many times since the phenomenon was named for Pliny, a Roman naturalist, who observed just such an event in 79 A.D. at Italy's Mount Vesuvius. During the 1902 eruption of Mount Pelée on the island of Martinique, the first detailed observation and recorded description of a nuee ardente (glowing cloud) associated with a plinian column was made. Over the ensuing years, field evidence has gradually accumulated on the nature of massive volcanic eruptions and production of the resulting ash deposits. These were poorly

Figure 6-1 Wineglass Welded Tuff *– This portion of the Wineglass Welded Tuff along the north caldera rim represents a collapsed part of the plinian column. After the ash stopped moving and settled, the deposit was still sufficiently hot so that the volcanic glass particles welded together into dense, hard rock. As a result, this exposure is often mistaken for a lava flow (Box 6-2 and Plate 9).*

understood until the 1960s, but in a breakthrough, USGS geologist Robert Smith proposed a model that tied the massive ash flows found worldwide to the processes that form calderas. Using his work and that of many others, geologists have identified hundreds of calderas based on field studies of ash flow deposits. These theories and associated models are critical to understanding the climactic event that created the Crater Lake caldera.

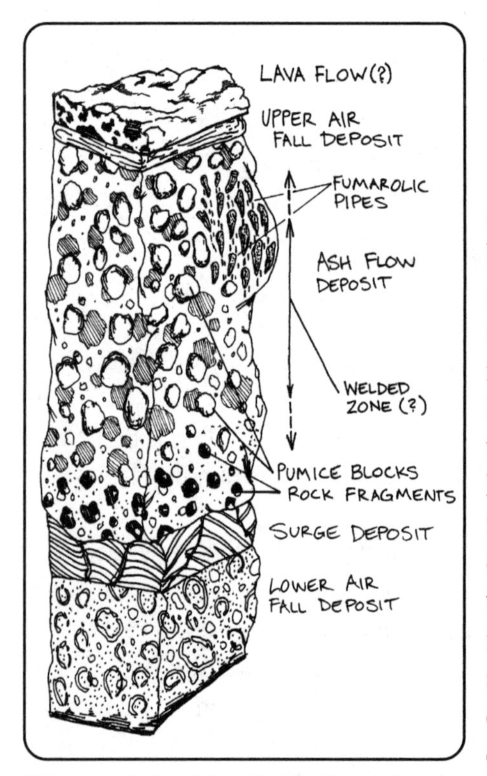

***Figure 6-2 Idealized Pyroclastic Section** – This sketch represents a schematic view of a complete eruption sequence. The lower air-fall is deposited first, and the upper air-fall (or lava flow) is last to form. The center represents the bulk of the erupted material, with interior portions that may become hot enough to fuse together, forming a welded zone (Figures 6-1).*

Air-Falls and Ash Flows

The way that pyroclastic material ejected during a large volcanic eruption behaves is extremely important in understanding how a caldera forms. Ash produced during an explosive eruption can result in several types of deposits – Williams recognized two of these in the Crater Lake materials. Today these are known as air-falls and ash flows. Only a few examples (Krakatau and Tambora) of major eruptions resulting in the formation of a large caldera have been observed in real time. However, a number of smaller events are documented. These observations, along with theoretical models, form the basis of our interpretation of how calderas form.

In recent decades it has been found that pyroclastic deposits in many parts of the world are related to caldera-forming eruptions. A complete section of such a deposit usually contains four or five distinct zones (Figure 6-2). The first to form, located at the bottom, consists of erupted materials – coarse-sized but lightweight pumice fragments – deposited by falling out of the air. This basal air-fall is often overlain by finer but more dense pumice particles, mineral crystals, and rock fragments, known as a surge deposit. Normally it is relatively thin and shows bedding (distinct layers). Next is the ash-flow section, which consists of the largest amount of erupted material and is usually the thickest. One peculiar feature of this section is the inverse depositional pattern – a tendency for larger particles to migrate toward the top. This characteristic occurs because the larger pumice blocks tend to "float" upward as the ash flow moves downslope before final deposition. A thin zone of fine air-fall ash accumulates slowly on top of the ash flow to create the fourth layer. In some places there is a final layer, a lava flow that caps the entire sequence.

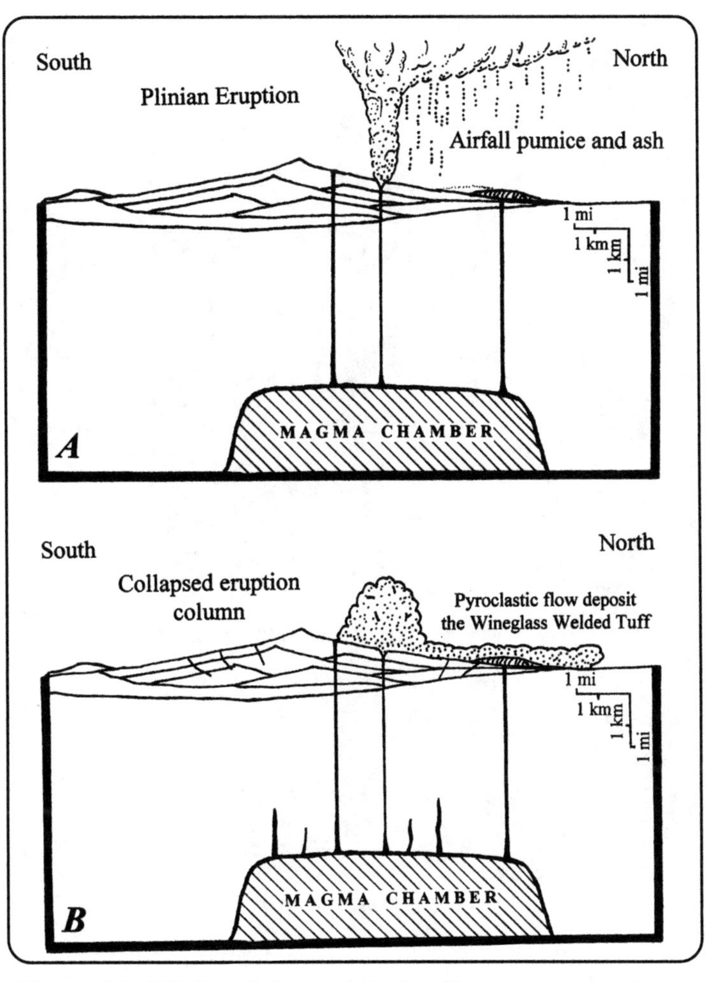

***Figure 6-3 Plinian Column** – As the climactic eruption began, a massive plinian column developed on the northeast side of Mount Mazama (A). The ejected materials were blown high into the atmosphere and carried to the northeast by prevailing winds. In (B) a portion of the column collapsed, resulting in the deposition of the Wineglass Welded Tuff as the caldera began to form. (After Klimasauskas, 2002.)*

Figure 6-3 is a sketch illustrating the way Bacon envisioned how the ash deposits formed during the violent climactic eruption of Mount Mazama. The first

***Figure 6-4 Mount St. Helens' Plinian Column** – Although Mount Mazama's column would appear similar, it would have been much larger. The amount of material ejected by Mount St. Helens was about 0.5 km^3 (0.25 mi^3) compared to about 50 km^3 (12 mi^3) during Mazama's climactic event.* (Photo of the May 1980 Eruption.)

air-fall deposit formed from a gas-rich vertical cloud (plinian column) rising to a height of several kilometers (miles) above the vent, depending on the pressure. Heat residing in the original magma is carried upward by the gas to reduce the cloud's overall density. This produces a convective effect that may rise many kilometers (miles) into the atmosphere, continuously expanding in the process. In time, as the gas pressure is relieved, a series of ground-hugging ash clouds will develop. These massive pyroclastic flows, often associated with composite volcanoes, are thought to occur when a portion of the plinian column collapses and the mixture of gas and solid particles plunges down the flanks of the cone. In ash flows a mixture of gas, pumice, mineral crystals, and rock fragments flows along the surface and may reach speeds ranging from 14 km/hr (9 mi/hr) to as high as 225 km/hr (140 mi/hr). Rapid movement of the subsequent ash flow(s) is due to the hot gases, which continue to expand and lift (or push) the solid material being transported downslope. Continuous mixing of these solids with larger pumice blocks "floating" upward prevents development of laminar bedding. In some cases the flows may be preceded by a blast of gas and finely suspended pyroclastic debris (the surge deposit). After the surge and ash flows have settled, a thin layer of air-fall gradually accumulates, representing the fine ash that was carried upward by the energy of the ash flows.

Not all of these zones in the ideal section can be seen in the exposures at Crater Lake. In examining the materials formed and deposited during Mount Mazama's final days, however, much of this theoretical model can be identified. Understanding the way pyroclastic material is erupted and deposited is the key to interpreting most of the volcanic deposits that makes up the terrain in Crater Lake National Park.

The Final Event

Recent fieldwork by Bacon and others have modified Williams' interpretation of the climactic eruption. Not far from the vent that produced the Cleetwood flow, a massive plinian column exploded to initiate the first part of the climactic eruption, known as the single-vent phase. It was similar to that of the 1980 eruption of Mount St. Helens (Figure 6-4) and Mount Pinatubo in the Philippines (1991), but much larger. Volcanic ash from this single vent was carried by prevailing winds for thousands of kilometers (miles) to the northeast. Fine material from Mazama has been found in western Canada and even in ice cores taken from Greenland glaciers. These deposits represent the bottom layer of the ideal section.

As pressure during this single-vent phase subsided, the summit region of Mazama started to sink along a circular pattern just inside where the caldera walls are today – the so-called ring-vent fractures. Speculation is that the upper portion settled like a piston

Box 6-1
Plinian Eruptions

A mixture of gas and solid particles, new volcanic materials, and older rock is discharged in a massive vertical column, often from a single vent (Figure 6-4). Velocities range between 100 and 400 m/s (325 and 1,350 ft/s) and volumes between 0.1 and 10 km^3 (0.024 - 24 mi^3). Due to these factors, the durations of plinian eruptions are limited – usually to a few hours or days. Hot, rapidly expanding gases provide the initial driving force, trapping and heating air at the margins. Although solid particles are mixed in with the gas and air, the column is much lighter than the surrounding air and may rise over 35 km (20 mi) in the atmosphere. Plinian eruptions are usually associated with silica-rich composite volcanoes.

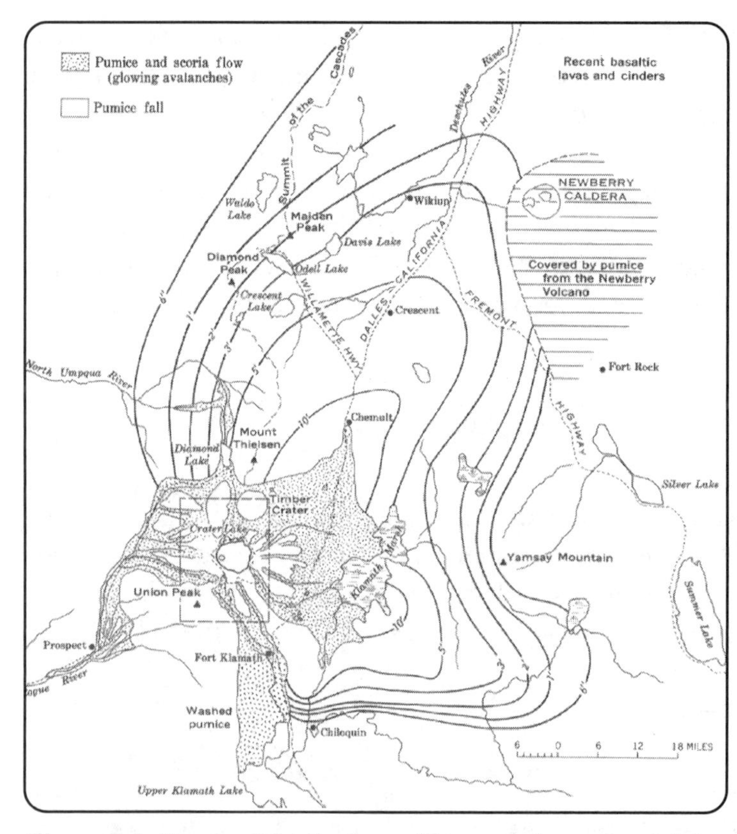

Figure 6-5 Pumice Distribution *– This map shows the vicinity of Crater Lake National Park. Contour lines represent air-fall (single-vent phase), while the stippled areas are ash flows (ring-vent phase). (After Williams, 1942, and Bacon.)*

into the interior space originally occupied by material that was being ejected from Mazama's magma chamber. This represents the beginning of the second part of the climactic event, known as the ring-vent phase. Volcanic materials continued to be erupted from the ring fractures, but now as massive ash flows (third layer of the ideal section) that moved down Mazama's remaining flanks in all directions. These hot materials filled the valleys and continued on for great distances away from the caldera.

The Single-Vent Phase

Most of the volcanic materials produced during the climactic event at Mount Mazama were ejected during the plinian column phase. The column formed from a single vent a little northeast of the present caldera center. Two units were produced in succession during the single-vent eruption, a climactic pumice fall (air-fall) and the Wineglass Welded Tuff. Initially pumice was thrust high into the atmosphere and carried away toward the northeast by prevailing winds. As the plinian column continued, the pressure driving the eruption was gradually reduced. In addition, the vent opening was enlarged by erosion or slumping, which also lowered the force driving the eruption. As a result, portions of the column forming its outer margins began to collapse. As this hot material settled to the surface, it moved downslope toward the north away from the vent as an ash-flow – forming the Wineglass Welded Tuff described by Williams. The single-vent phase ended suddenly, as the Wineglass was being deposited, with the inception of caldera collapse.

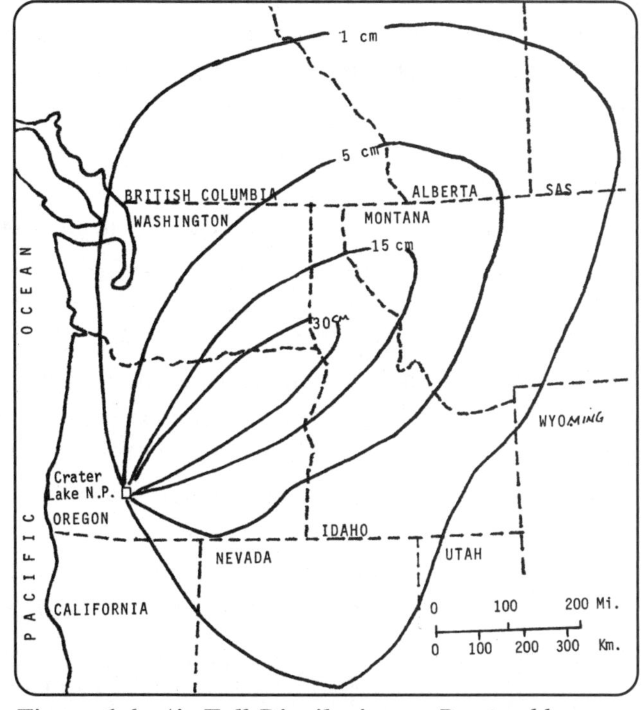

Figure 6-6 Air-Fall Distribution – *Pumice blown out during the single-vent phase of the climactic eruption has been found across the northwestern United States and western Canada (Plate 12). (After Lidstrom, 1972.)*

Air -Fall Unit – Deposits of this part of the climactic eruption are found as beds distributed over a huge area of the northwestern United States and western Canada. Nearer the caldera these deposits consist of bedded pumice lapilli and blocks, some up to 60 cm (24 in.) across. In addition, lithic (rock) fragments more than 1m (3^+ ft) are also included in these bedded materials. These ejecta reach thicknesses of about 20 m (~70 ft), but this likely represents a minimum as pyroclastic flows during the ring-vent phase probably removed some of the loose surface materials. Also, it must be noted that air-fall deposits are not usually seen on the surface near the caldera because later ash flows have buried them.

Farther from the caldera, the climactic pumice deposits are tephra, and beds gradually decrease in thickness with distance (Figure 6-5). At Newberry Crater, some 110 km (~70 mi) northeast, Mazama's climactic (single-vent phase) ash is 50 cm (~20 in.) thick, and even at 1,200 km (~750 mi) distance in Saskatchewan, Canada, the layer is still over 1 cm (~0.4 in.) (Figure 6-6). On some steeper terranes, pumice has been washed into depressions to form deposits up to 15 m

(50 ft) thick. Williams notes that the regularity of the deposits suggests they were laid down quickly. This infers the climactic eruption lasted a relatively short time. As expected with wind-deposited materials, the particle size becomes smaller with distance from the source, and this is the case with the air-fall from Mazama's single-vent phase.

Most of the new ejecta, juvenile material, produced during the single-vent phase was rhyodacite with a composition between 70 and 71% SiO_2. There was a small amount of phenocryst-rich pumice and andesitic scoria mixed in, including rare scoria bombs as large as 90 cm (35 in.) across. Some of this air-fall was hot enough near the vent that a small amount of welding occurred. As might be expected, Mazama ash has been identified in exposures of strata at many other locations throughout the Pacific Northwest. Because individual volcanic eruptions have unique chemical and mineralogical characteristics, deposits of ejecta can be traced to that event. And, since such deposits are emplaced in a short period of time, this permits an exact date to be assigned to that layer regardless of its location. This technique provides a valuable time reference in a sequence of rocks and is known as a stratigraphic marker.

As coarse ash fell on the still hot Cleetwood lava, some was welded and reddened due to oxidation of iron in the volcanic glass. In a matter of weeks (or perhaps just days), the summit of Mazama would settle into the interior, and the still fluid middle portion of the Cleetwood flow would sag into the newly formed caldera. In the meantime, another significant feature was forming – the Wineglass Welded Tuff. This distinct layer, named for its interesting shape on the caldera wall, can be seen along most of the northern rim of the caldera.

Box 6-2
Welding

This process is caused by the compaction of hot pyroclastic material due to its own weight. It occurs when individual particles slowly deform and stick together as they accumulate. Factors involved are the weight of the overlying material, amount and retention of heat by the particles (related to particle size, amount of time in flight, and initial temperature), and rate of accumulation. If deposition is rapid and initial temperatures high, particles may become highly welded – that is, they strongly adhere, appear as a solid, and do not break apart easily. With slower deposition and lower temperatures, deposits stick together but are more easily broken apart. At the very high end of this spectrum (densely welded), welded volcanic materials may appear as a lava flow (Figure 6-1).

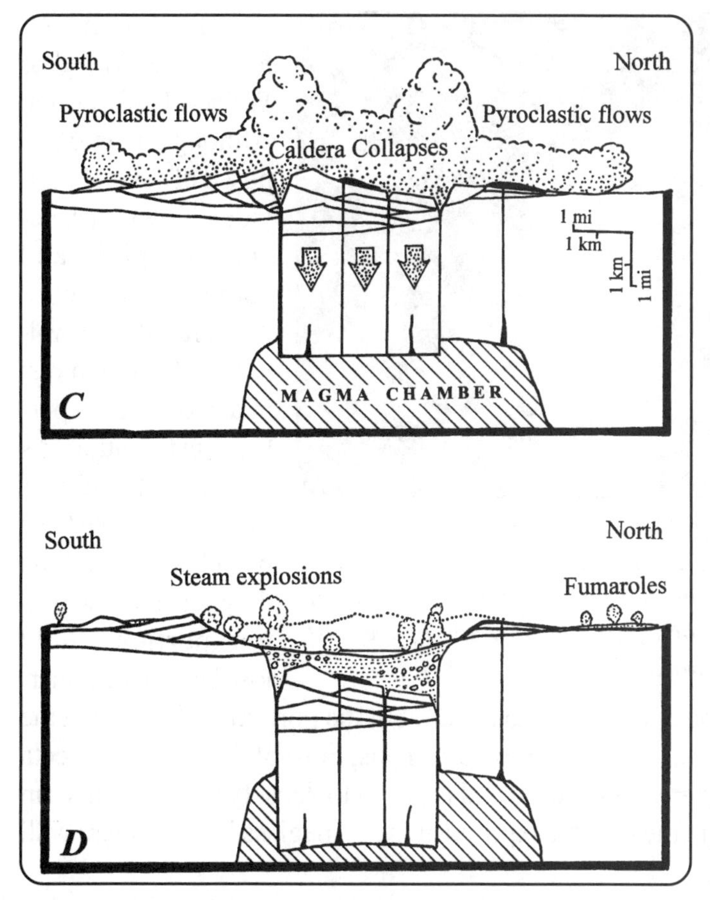

Figure 6-7 Ash Flows *– During the ring-vent phase of the climactic eruption, massive amounts of pyroclastic debris charged down the slopes of Mount Mazama (C). These flows covered the slopes on all sides of the volcano, filling glacial valleys and continuing out onto flat regions beyond. In (D) the summit has foundered into the interior of the mountain, followed by a period of volcanism. (After Klimasauskas, 2002.)*

Wineglass Welded Tuff – With enlargement of the vent and reduction of pressure, marginal portions of the plinian column began to collapse. The material cascaded back to the surface and moved downslope as an ash flow toward the north, away from the vent. It followed valleys and is now found thickest in topographic lows and thin or absent on higher areas. This aspect can be easily seen on the caldera walls where it occurs between Pumice Point and Skell Head. Exposures of the Wineglass Welded Tuff are 3^+ m (~ 10^+ ft) thick in some locations and can be found as far as 11 km (~7 mi) east of the caldera. Based on its distribution, its volume is greater than would be expected from exposures in the caldera wall. Because much of it is buried beneath later deposits, the volume cannot be accurately determined. However, it is likely several cubic kilometers (one cubic mile?). Also, its distribution is a critical factor in determining the location of the initial vent (single-vent phase) and topography of Mount Mazama. It was bounded by the Llao Rock and Redcloud flows on the west and east, and Mazama's high topography to the south.

***Figure 6-8 Rim Village Deposit** – Thick deposits of ring-vent phase pyroclastic ejecta seen here can be found in many locations around the caldera rim. The Sinnott Memorial Overlook (far right) provides one of the best views of the interior walls of the caldera.*

The material forming the Wineglass Welded Tuff was hot. Although it is composed of several individual flows, they occurred rapidly and cooled as one unit. These raced down valleys and came to rest in a short time – perhaps just minutes. As a result, little of their initial heat would have been lost by the time they settled. The bottom portion would lose heat to the underlying material and the upper zones to the air, but the portion of the flow in the middle would remain at a high temperature. Thus, most of this volcanic ejecta experienced some degree of welding and is partly to densely welded. This can readily be observed in the section exposed at the Wineglass in the caldera wall (Figure 6-1).

The Wineglass Welded Tuff's composition is identical to that of the air-fall, produced by the plinian column, that preceded and underlies it. Even the rock fragments are similar. This argues for the idea that a portion of the plinian column collapsed as the vent enlarged due to erosion or slumping. It also provides evidence that the caldera started settling at about the same time to begin the ring-vent phase. Evidence for this timing is found where the WineglassWelded Tuff grades upward into materials of the ash flows produced during the ring-vent phase.

The Ring-Vent Phase

As the summit of Mazama sank to form the caldera, the nature of the eruption changed dramatically. Instead of being blown high into the air, ejected volcanic material was vented from fractures in an oval pattern that allowed the upper part of the volcano to collapse (Figure 6-7). The highly gas-charged materials

__Figure 6-9 Ash-Flow Bedforms__ – These are clearly outlined by snow patterns on pumice fields near Rim Drive on the north slopes of Mazama. Red Cone can be seen on the left and Mt. Thielsen's glaciated core in the distance.

flowed out of these new vents ringing the flanks of Mazama and down the surrounding valleys. There must have been many high erupting columns around the subsiding block to explain the mobility of the pyrocastic flows and lack of dense welding. They were hot and moved rapidly – perhaps hundreds of kilometers per hour (miles per hour). All valleys were filled, some to a depth of nearly 100 m (~300 ft).

Deposits of the ring-vent phase can be divided into three zones (or facies): proximal, medial, and distal, depending on the distance from the caldera each was deposited. Evidence from the study of these deposits is important to interpreting the final days (or hours) of Mazama's climactic eruption.

Proximal Facies – This part of the ring-vent material is found around the caldera rim, on the near slopes of Mazama, and in the heads of some valleys (Figure 6-8). Deposits are up to 20 m (65 ft) thick and have a loose breccia-like appearance. This may have been the reason that Williams interpreted some as glacial till. They have a complex mixture of compositions and sizes that include coarse material like bombs and lithic fragments, scoria, and even sizes down to fine pumice. However, being deposited by rapidly moving flows, much of the fine

***Figure 6-10 Pinnacles** – Hot ash flows produced during the ring-vent phase of the climactic eruptions filled existing valleys. Gas escaping from fumarole vents caused the surrounding material to become resistant to later erosion. This classic view is located in Sand Creek Canyon at the end of Pinnacles Road.*

material was carried away. In some places these were hot enough to be altered by fumarolic activity as trapped gases escaped. As the flows moved across the surface, bedforms resembling sand dunes formed in some locations, but on a much larger scale (Figure 6-9). Like the Wineglass Welded Tuff, proximal deposits are composed of many individual flows that succeeded one another. These deposits range between 2 and 11 km (1.3 – 7 mi) out beyond the caldera, where they grade into the medial facies.

Medial Facies – Deposits in this portion of the ring-vent phase are perhaps best known for they include the spectacular compositionally zoned ejecta. The pinnacles in Wheeler, Sand, and Annie Creek canyons are examples used to illustrate many geology textbooks and publications (Figure 6-10 and Plate 14). In these locations and many other valleys buried by the ring-vent flows, three zones are represented: (1) a lower light-colored non-welded rhyodacite pumice, (2) the middle (partly welded in some canyons as widely spaced columnar joints) grey-colored layer that is also rhyodacite, and (3) the upper dark-grey andesitic deposit. The chemistry of these materials provides evidence of the zoned nature of Mazama's magma chamber prior to the climactic eruption. Pumice and scoria blocks here have a range of SiO_2 between 71% in the lower portion to 54% near the top.

Figure 6-11 Pinnacle Detail *– This close-up of a pinnacle illustrates the resistant portion created by hot gases venting through a fumarole.*

The hypothesis here is that magma was converted to pyroclastic ejecta starting at the top of the chamber and progressively drew material from deeper portions as the eruption continued. Thus, the silica-rich upper portions of the chamber were deposited first and now occur beneath material originally lower in the chamber. Thus these deposits are "upside down." Once the flows stopped moving and settled, the hot gases they contained began to percolate up through the ejecta. At localized vents, fumaroles developed where gases migrated to the surface. In addition, adjacent material was heated, and dissolved minerals from escaping gases were deposited (vapor-phase crystallization) in the spaces between particles that hardened the vent walls. Subse-

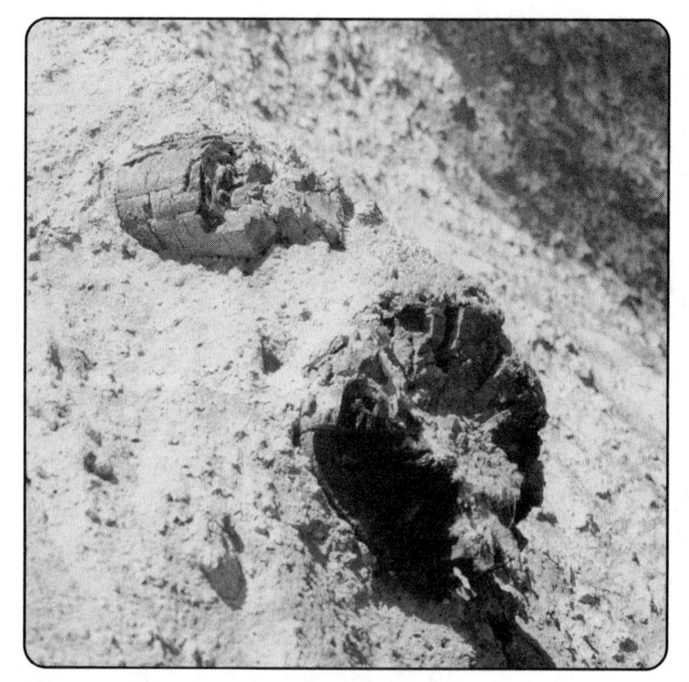

Figure 6-12 Carbonized Logs – *Examples like this tree buried in the ring-vent phase ash flows have been found in many Mazama climactic eruption deposits.*

quent erosion by streams reestablishing their valleys removed the less resistant materials between the fumarolic pipes, leaving the latter standing as prominent pinnacles (Figure 6-11). These are best observed along Pinnacles Road and in Annie Creek Canyon (Figure 7-12).

Distal Facies – Beyond about 20 km (13 mi) and as far as 60 km (40 mi) from the caldera, the ring-vent phase material is silicic pumice. It is usually nonwelded as temperatures were lower, although this material still contained significant exsolving hot gases. This produced secondary explosion craters when the flows came into contact with water. It also produced an interesting phenomenon where larger pumice blocks tended to "float" to the top of the deposit. These are the deposits where carbonized wood (buried trees) are often found that have been used to date the climactic eruption (Figure 6-12). As these flows moved along, a cloud of fine, well-sorted ash formed and settled in layers on top, in some areas extending beyond the farthest reaches of the distal material.

Mazama's Magma Chamber

Working out the nature of Mazama's magma chamber is a difficult problem since no direct real-time observations are possible. The following description is based on many factors and draws heavily on theories of magma behavior below the Earth's surface, field relations and mineralogy of erupted materials, and modeling of volcanic systems. While it represents a reasonable explanation of how the

Crater Lake caldera was created, to a degree it's speculation.

Development of the magma chamber that ultimately produced the caldera resulted from a long period of heating associated with Mazama's magma system (Figure 6-13). Prior to about 70,000 years ago (Table 4-1), there was little silica-rich magma involved in eruptive ejecta at Mount Mazama. Then, starting with the dacite dome(s) that formed on Mazama's southern slopes, silica-rich magma became the typical composition of erupted ejecta. The precursor events, Redcloud, Grouse Hill, Llao Rock, and Cleetwood produced rhyodacite and clearly illustrate this change in composition. It appears that over this period a fundamental change occurred in the composition of Mazama's magma chamber, an important factor leading up to the climactic event that occurred 7,700 years ago.

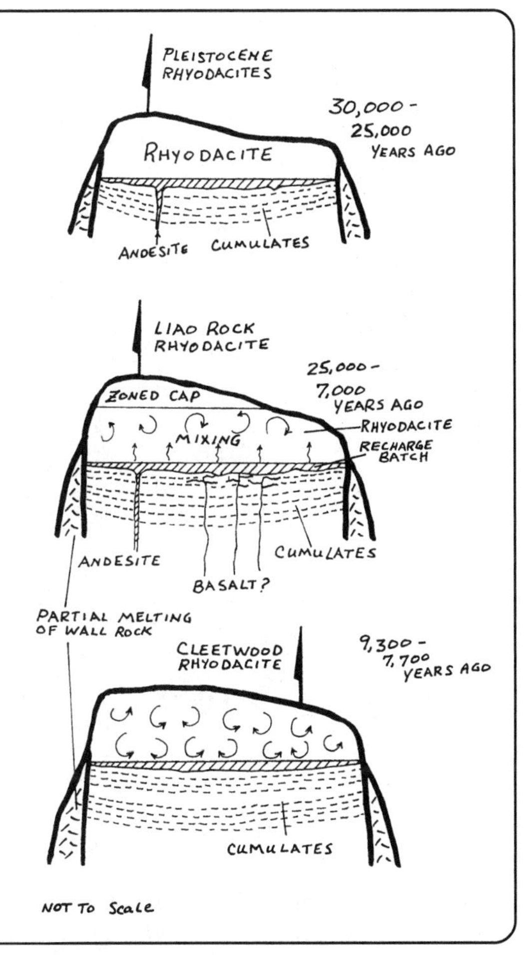

***Figure 6-13 Magma Chamber** – Mazama's magma chamber developed over a long time, gradually changing in composition. The pre-climactic eruptions occurred during this period. (After Bacon and Druitt, 1988.)*

At the time of its cataclysmic eruption, Mazama's magma chamber contained some 50 km^3 (12 mi^3) of rhyodacite, which accumulated over a period of time, but at a rapid rate compared to other volcanoes. The chamber was composed of a series of layers, with the rhyodacite forming the upper level. It was underlain by an andesite zone with more mafic magma at the bottom. These layers formed what is known as a zoned magma chamber. Modeling and experimental evidence imply that the parent magmas that fed into the chamber were mainly andesite in composition. The concept of a parent magma holds that the

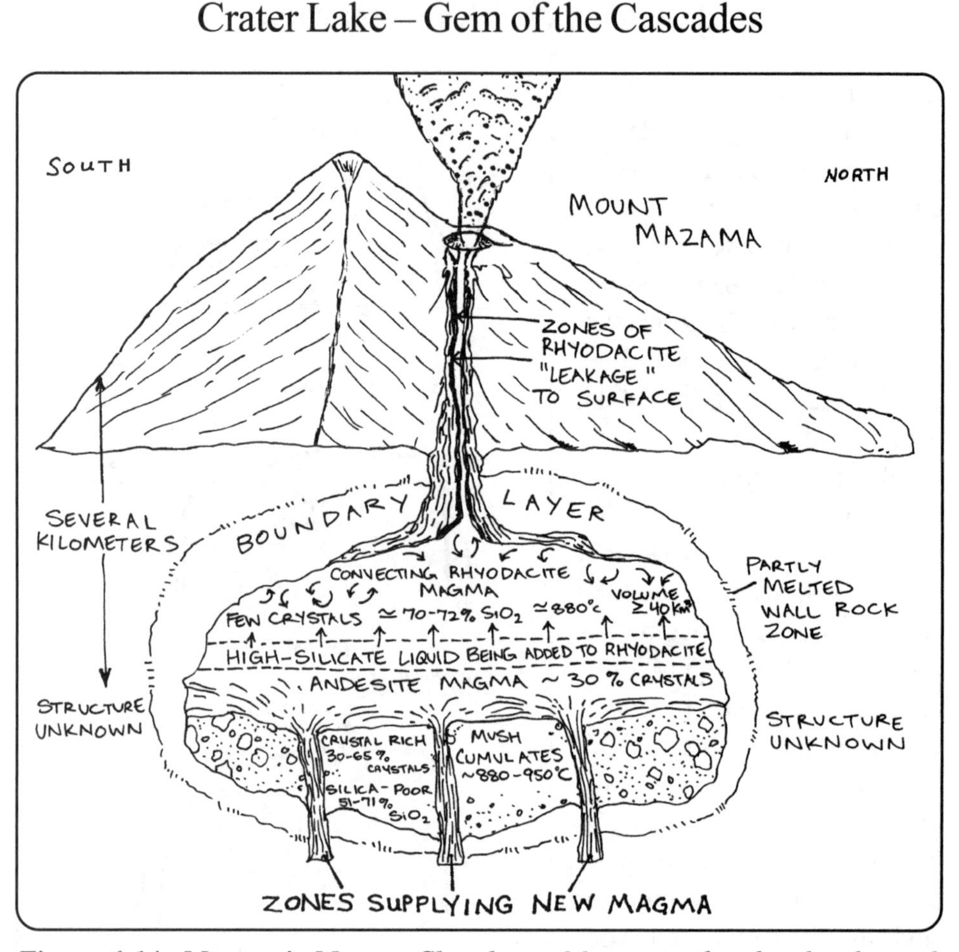

Figure 6-14 Mazama's Magma Chamber – *Magma in the chamber beneath Mount Mazama just before the climactic eruption was a complex mix of compositions. This condition is thought to have caused the massive climactic event, even though the exact reason is not well understood. This sketch is only schematic, and the actual vents were north of the highest part of Mazama's edifice. (After Bacon and Druitt, 1988.)*

liquid rock entering a magma chamber is the source from which other magmas (or igneous rocks) are derived. A number of processes may modify this initial magma to produce a range of different compositions of "new" fluids and solids (crystals). The lavas erupted by the monogenetic and shield volcanoes in the park are also of interest here as they may represent the range of compositions for parent magmas that fed Mazama's magma chamber.

The rhyodacite had an SiO_2 range between 70 and 72%, while the underlying andesite (parent material) and accumulated crystals were 56 to 62% SiO_2.

Figure 6-15 Lag Deposits – *These bombs may represent the mixing of different portions of Mazama's magma chamber. During the climactic eruption, ejecta was carried across the surface. Larger and heavier materials dropped out, while the fines continued to move. Similar bombs are scattered across the Pumice Desert in the northern portion of the park.*

Based on detailed study of the materials deposited during the climactic event, Bacon and others have worked out a physical model for the evolution and nature of Mazama's magma chamber just before the eruption (Figure 6-14). Fundamental to their explanation is the repeated recharge of the chamber by a parent magma in association with crystal fractionation, crystal accumulations, wall rock assimilation, and magma mixing.

In this interpretation, the essential aspect is formation of a convecting rhyodacite magma near the top of the chamber, produced by rising low-density fluid. This fluid was "left over" from a process where certain minerals crystallize and are removed due to their higher density than the remaining melt. Such minerals are rich in iron and magnesium and begin to form at temperatures above those of the silica-rich magma collecting in the upper zone. Based on evidence from climactic eruption pumice, temperatures in the upper portions of the chamber were nearly 900°C (1,650°F). Below, however, in the cumulate layers temperatures were at least 950°C (1,750°F) or higher.

An influx of andesite and basaltic magma was repeatedly added to Mazama's chamber, a process called recharge. Although it is uncertain how this worked, it must have occurred in the underlying andesite layer, allowing the residual fluid melt to rise while the dense crystals settled to form the cumulate (mafic) layers below. Apparently, as the andesite magma cools, crystals form, and the remaining liquid separates and rises to the top of the recharge layer, eventually mixing in with the silicic zone above. In addition, the andesite layer's higher temperature helped heat

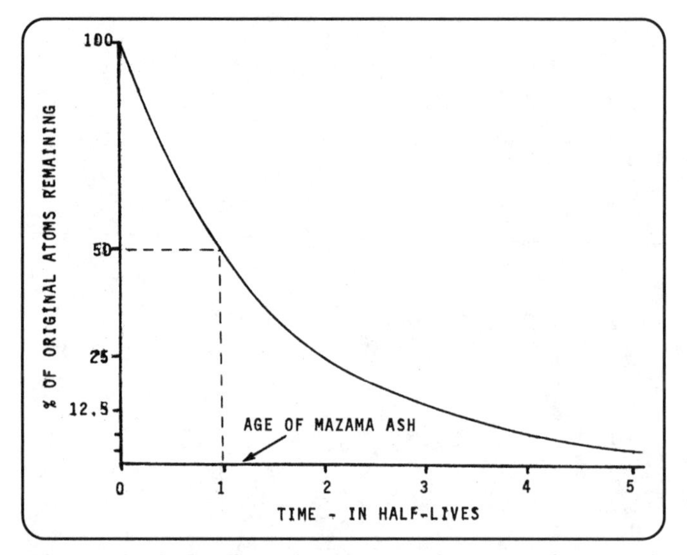

***Figure 6-16 Radiocarbon Decay Curve** – This graph illustrates the relationship between the age of a material and amount of radiocarbon it contains. Using organic material found in Mazama's climactic ash establishes when the caldera formed – about 7,700 years ago.*

the rhyodacite above, providing energy to keep that portion of the magma mixed up. This behavior, a phenomenon known as convection, is similar to what happens in a boiling pot of water. Periodically, new andesite magma was injected into the middle zone to keep the system going and gradually increased the volume of silica-rich magma in the upper layer. Apparently this situation continued for several thousand years, increasing the volume of rhyodacite and gas content and building to the climactic event. There was some "leaking," however, which resulted in the Redcloud, Grouse Hill, and Llao Rock eruptions.

Both preclimactic events that happened just before the climactic eruption, the Llao Rock and Cleetwood flows, and the single-vent phase tapped the upper rhyodacite portion of Mazama's chamber. Evidence for this is that the Cleetwood pumice and flow and the climactic pumice (single-vent phase) have nearly identical compositions. This suggests vigorous convection in the upper zone of the magma chamber. The climactic magma was gas-rich and produced the initial plinian column. Later, the ring-vent phase also contained gases that percolated to the surface through fumaroles in the valley-filling materials. Evidence suggests that gas content consisted primarily of dissolved water (H_2O at about 4–5%), sulfur gases, and small amounts of chlorine (Cl) and fluorine (F) along with other gases commonly associated with explosive volcanic eruptions. During the ring-vent phase, all portions of the chamber were ejected sequentially, resulting in the compositions observed in the eroded valleys today. Near the end of the climactic eruption, the lower crystal-rich cumulative zones of the chamber were tapped to produce the dark scoria bombs and blocks seen in many locations on the surface today. This is apparent in veneer deposits of crystal-rich scoria and bombs near the caldera rim that represent the final products of the climactic event.

In addition to the andesitic magma that is thought to have recharged Mazama's chamber, there is some evidence that a more mafic (basalt?) magma was also present, most likely at or near the bottom of the cumulate layers. Certain of the scoria blocks have compositions that lead to this interpretation (Figure 6-15).

This is a much simplified version of Mazama's magma chamber and what actually happened during the final massive eruption. Numerous other factors must have been involved: the specific compositions of various magma zones, interaction with the wall rock, temperatures of various zones and others. It must be remembered that the sketch illustrating the magma chamber (Figure 6-14) is merely a schematic of the general idea of what the actual conditions were thought to be at the time of the climactic event.

Dating the Crater Lake Caldera

Dating events at Crater Lake is one of the most fascinating studies associated with the geology of the park. In previous chapters, a number of dates have been given for various rocks and materials produced during Mazama's construction. These were based on a number of techniques, but for most of them radiocarbon dating was not employed. This goes back to the nature of radioactive decay and the half-life of the carbon used in this method (Box 4-2). With a half-life of 5,700 years, material older than about 50,000 years has lost most of its radiocarbon. Since the climactic event occurred some 7,700 years ago, however, carbon dating is a good tool to date the formation of the caldera (Figure 6-16).

Because the climactic eruption and resulting air-falls and ash flows occurred over a brief period, anything living at the time and buried would have a radiocarbon age equal to Mazama's collapse. Prior to this massive eruption, the surrounding landscape was forested. In most cases the trees covered by Mazama ejecta were killed, decayed, and disappeared. Under certain circumstances, however, trees were converted to charcoal and preserved. This occurred where they were buried in hot pumice or ash. Under these conditions, trees would not burn, due to the lack of oxygen, so the volatiles in the wood were driven off as gas, leaving only charcoal. This process does not affect the radioactive carbon which then begins to decay in the usual way. Thus when a charcoal tree is found, buried in Mazama material, samples can be dated. It should be noted, however, that close to Mazama, pyroclastic flows thoroughly incinerated (vaporized) the forest. As a consequence, carbonized logs tend to be found some distance from the caldera.

Determining a date using carbon is an elaborate and sophisticated process. As a result, there is some confusion associated with the radiocarbon dates of materials found in Mazama climactic materials. Initial findings were listed at about 6,850 years BP. This date was widely used for many years and will be found in older references (prior to about 1992). In recent years, however, the current value of 7,700 years BP is the accepted date for the collapse of Mount Mazama to form the Crater Lake caldera. This "new" date is simply the calibrated age where dendrochronology (tree-ring chronology) has been used to calibrate Carbon 14 dates. This is necessary because the production rate of Carbon 14 in the atmosphere has not been constant.

Additional Reading

Klimasauskas, E., Bacon, C., and Alexander, J., 2002, "Mount Mazama and Crater Lake: Growth and Destruction of a Cascade Volcano", USGS Fact Sheet 092-02. Describes the climactic eruption that resulted in the collapse of Mount Mazama's summit area.

Last Days at Mount Mazama

Chapter 7
Beneath Crater Lake

With the collapse of Mount Mazama's summit area and formation of the Crater Lake caldera, most of the geological drama was over. However, a number of important events have occurred in the 7,700 years since the basin formed. Some of these are volcanic in nature, representing the dying volcanic activity of Mazama, while others reflect the adjustment to a new topographic feature – the sudden presence of the Crater Lake caldera.

One of the first research efforts at Crater Lake was an attempt to determine its depth in 1886 (Figure 1-4 and page 5). Continuously improved techniques for measuring depths, collecting materials, and making direct observations of the lake's floor since then have been the focus of many projects. Additional soundings using wire gear were made in 1938 – 40 by park staff, and in 1959 the U.S. Coast and Geodetic Survey completed some 4,000 acoustic depth measurements. In the early 1980s samples were collected and a remotely operated vehicle (ROV) was used to photograph portions of the caldera floor. Exploration using a manned submersible (Deep Rover, Figure 7-2) was conducted during the summers of 1988 and 1989. Finally, in August 2000, a team composed of members from the U.S. Geological Survey, University of New Hampshire, C and C Technologies, and the National Park Service conducted an extensive mapping of Crater Lake's floor.

Figure 7-1 Wizard Island *– The Wizard Island cinder cone rests on its underlying lava platform and is surrounded by lava flows. Those extending toward the western caldera wall form the narrow area known as Skell Channel (lower right), the shallowest region of Crater Lake.*

Figure 7-2 Deep Rover *– This one-person submersible was used to explore the caldera floor during 47 dives over two summer seasons (1988–89). It was able to function in the deepest portions of Crater Lake, which allowed for taking temperature measurements, filming the floor, and collecting water samples, rocks, and biological specimens.* (NPS image.)

This most recent effort used high-resolution multibeam echo sounder and GPS technology to create a detailed map of the caldera's interior surface below lake level (Figures 7-3 and 7-4). This survey collected over 16 million soundings, and the resulting data portrays the bottom of Crater Lake at a spatial resolution of 2 m (~7 ft). Using all the data gathered from these surveys, Bacon and others have interpreted the geology and sequence of events following the collapse of Mount Mazama. In this chapter, these post-collapse geological activities at Crater Lake National Park will be examined.

Post-Collapse Activity

Even though Mount Mazama's summit disappeared into its interior, volcanic activity continued for some time. All evidence indicates that post-collapse eruptions were confined to the interior of the caldera. Rocks outside the caldera are part of the old mountain, resulting from activity prior to the collapse, or material deposited during the climactic event. Based on the data gathered to date, but primarily in 2000, several volcanic features have been identified on the floor of the Crater Lake caldera. The largest, and one of only two visible above Crater Lake's surface, is Wizard Island (Figure 7-1). Immediately east of the Wizard Island platform of lava flows, and resting on its slopes, is a rhyodacite dome. A bit farther east is the central platform. A final prominent volcanic feature on the caldera floor is Merriam Cone, about 3 km (2 mi) northeast of the central platform. Much of the rest of the floor is relatively flat, including the East Basin, Northwest Basin, and Southwest Basin. The small edifice of Phantom Ship, the other feature that can be

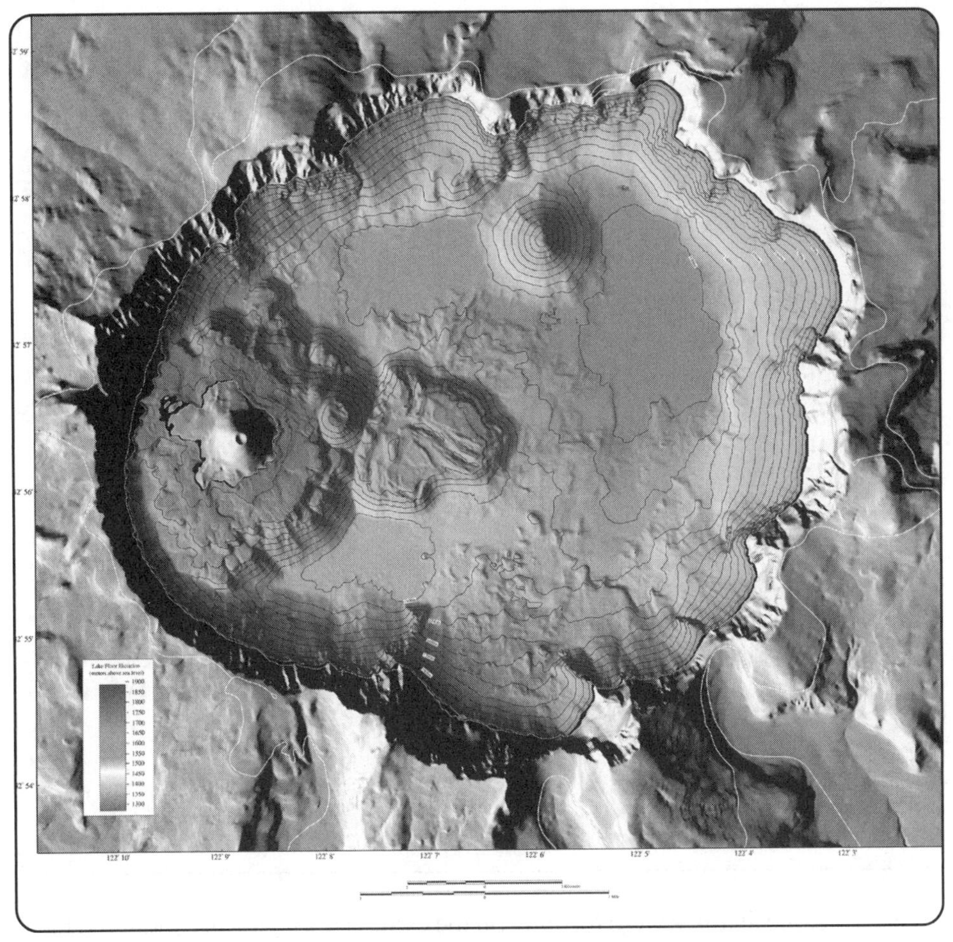

Figure 7-3 Caldera Floor *– This shaded relief map representing Crater Lake's floor was produced from the multibeam echo-sounding data. It illustrates the prominent volcanic and mass wastage features formed following the collapse of Mount Mazama some 7,700 years ago.* (USGS image.)

seen above lake level, is located along the southeast portion of the caldera wall and is composed of much older rocks (Figure 4-11 and Plate 6). During and following the collapse, significant portions of the exposed wall above lake level were overly steep, failed, and slid onto the caldera floor as landslides and debris avalanches. Table 7-1 summarizes the portion of the lake's floor covered by these features along with their volumes.

Estimates of how long it took Crater Lake to fill following the collapse vary greatly. The shortest is between 200 to 300 years to fill to its present elevation, while others range up to 700 years or more. During this period of rising water

Table 7-1
Caldera Features
Areas and Volumes*

Feature	Area km² (mi²)	%	Volume km³ (mi³)
Volcanic (postcaldera andesite)			
Wizard Island (above water)	1.22 (0.46)	2.3	0.06 (>0.01)
Wizard Island (below water)	7.75 (3.02)	14.5	2.56 (0.61)
Central platform edifice	3.19 (1.24)	6.0	0.76 (0.18)
Central platform lava fields	4.89 (1.91)	9.2	0.30 (0.07)
Merriam Cone	2.38 (0.93)	4.5	0.34 (0.08)
East Basin lavas	0.87 (0.34)	1.6	0.03 (<0.01)
Totals	20.30 (7.91)	38.1**	4.05 (0.96)
Rhyodacite dome	0.61 (0.24)		0.07 (0.02)
Debris-avalanche deposits			
Chaski Bay deposits	4.59 (1.12)		0.22 (0.05)
Others***	4.69 (1.83)		0.11 (0.03)

* Values have been rounded using 1 km^2 = 0.39 mi^2 and 1 km^3 = 0.238 mi^3.
** Balance of area is represented by the three basins, debris-avalanche deposits, and caldera wall.
*** Includes deposits below Danger Bay, Cloudcap, Grotto Cove, Steel Bay, Llao Bay, and Eagle Point.
Based on a total area of 53.4 km^2 (20.83 mi^2), after Bacon, etal, 2002.

levels, volcanic features were forming on the caldera floor. Former shoreline locations and other features that formed as the lake level rose have been recognized. One later addition was the rhyodacite dome, which appears to have formed underwater some 2,500 years after the collapse, most likely representing the last volcanic event in the park. Details of Crater Lake's filling along with the physical characteristics of its water are the topic of Chapter 8.

A common feature of large calderas is a circular fracture system located just inside the caldera walls (Figure 7-4). This zone, known as a ring fracture, represents the region outlining the plug of a volcano that subsided into the underlying chamber like a piston. Consequently it is usually the site of pyroclastic eruptions and dacite dome emplacement following the cataclysmic collapse of the volcanic surface that creates a caldera. Both Merriam Cone and Wizard Island are located approximately where a ring fracture should exist, if one were present. Before the early 1980s, there was little evidence to suggest the presence of a ring fracture.

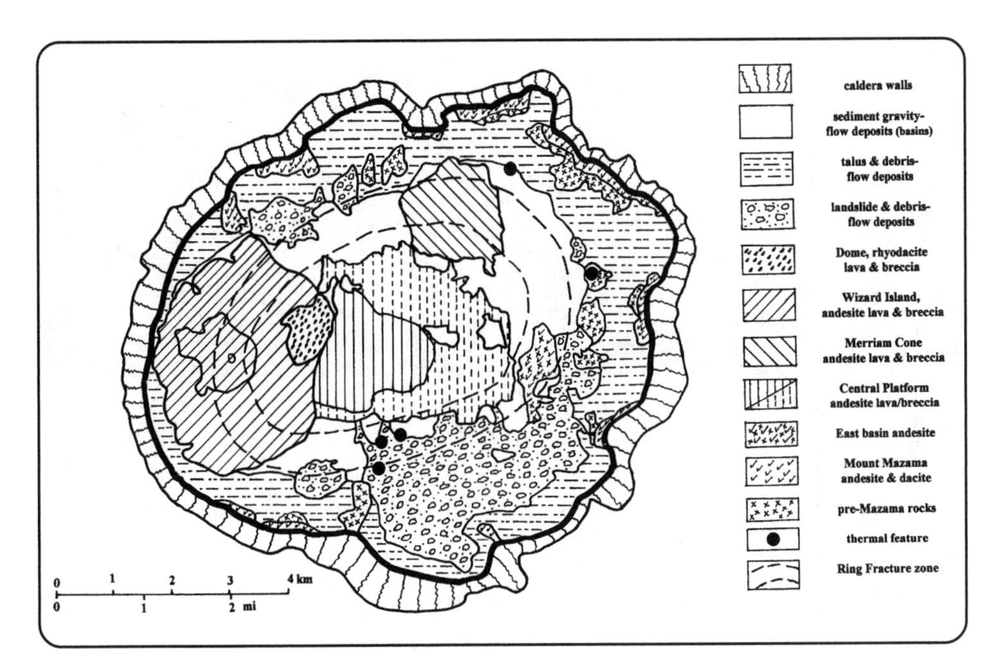

Figure 7-4 Geologic Map of Caldera Floor *– This generalized map of Crater Lake's floor shows the location and relationships between the post-collapse volcanic and mass wastage features. It is based on the shaded relief map in Figure 7-3. (After Bacon, 2002.)*

However, evidence for a ring fracture system is provided by the location of warm water vents, other thermal features, and bacterial mats.

Volcanic Features

Wizard Island – Perhaps the most spectacular cinder cone in the world rests along the western margin of Crater Lake. One story credits William Steel with naming Wizard Island because he thought it was shaped like a wizard's hat. Certainly, one of its most striking characteristics is the symmetry of the upper portion of the island. However, the Wizard Island platform is much more extensive than just the cinder cone. The visible portion above the lake is actually a small part of this post-climactic eruption feature. The cinder cone is associated with a complex of lava flows forming an underlying and unseen platform. The Wizard Island complex is by far the largest feature on the caldera floor (Figure 7-5).

The Wizard Island summit rises 750 m (2,500 ft) above its base. The area at its base is 7.75 km^2 (3.02 mi^2) with a volume of some 2.6 km^3 (0.6 mi^3). Most of

Figure 7-5 View of Caldera Floor *– This oblique view looking southeast across the basin was generated from the multibeam echo data. It shows the massive nature of the Wizard Island platform and cinder cone along the western margin of the caldera. The distance across the bottom of the image is about 6 km (3.7 mi). (USGS image.)*

this feature is a broad lava flow field below the lake surface that continues upward into flows and air-fall materials that compose the visible portion. Some of these flows lap up against the lower caldera wall to the west of the cone where active talus slopes lie atop them. This produces the shallow portion of the floor, known as Skell Channel, between the shoreline and the flows visible along the island's western margin (Plate 10 and Figure 7-1).

Figure 7-6 Wizard Island Platform *– This view is looking southwest toward The Watchman and Hillman Peak along the caldera rim. Notice the well-developed bench (drowned beach line) on the slope of the platform that drops off toward the East Basin. (USGS image.)*

Evidence suggests that Wizard Island formed as the lake was filling. Slopes drop quickly at angles of 30 degrees or more into deeper portions of the caldera floor. Certain features, known as benches, represent drowned lava flow surfaces (Figure 7-6). They indicate the location of a former lake level, marked by a break in slope. During eruptions, hot lava broke apart as it encountered the water, creating debris that formed the steep slopes below the benches composed of breccia. It appears these benches formed as lava flows reached the lake shoreline as the basin was being filled. Also, narrow submerged beaches were found along the northeast side of Wizard Island. These shallow features, less than 40 m (125 ft) deep, were formed by wave action when the lake level was lower and represent a pause in the eruptions. Indeed, they help constrain rates of a lake filling model and were used to put approximate dates on the eruption of Wizard Island and the central platform volcanoes. Some lava flows, however, remained intact to form ridges as they moved downslope underwater.

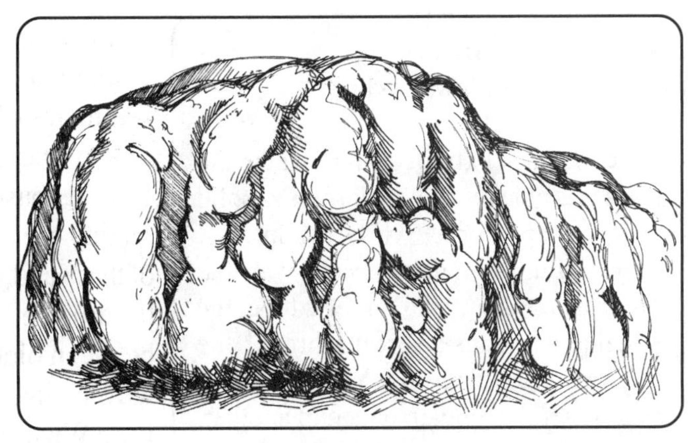

***Figure 7-7 Pillow Lava** – This sketch illustrates lava flowing over a slope to form pillows as they typically occur when basaltic flows encounter water. A similar feature may have occurred on Wizard Island and the central platform. With the much more viscous rhyodacites involved, the features formed would be larger and less well defined. (After Sigurdsson, 2000.)*

Lavas collected from these submerged portions of Wizard Island are angular fragments. In some places pillow-like structures are found with compositions similar to the lava flows exposed above water. Pillow lavas are direct evidence that fluid lava was cooled and solidified in water (Figure 7-7 and Box 7-1). In places, drowned subaerial lava flows appear to run from the base of the cinder cone down to the highest shoreline beach, suggesting that Wizard Island's eruptions had ceased before the lake filled to its present level.

Another major finding from the lake floor studies, related to filling of the lake, is a submerged wave-cut platform along some portions of the caldera wall. This indicates that the lake has been at its present level for a long time. It also is evidence supporting the concept that Crater Lake's level is regulated by a hypothesized aquifer below the lake near Roundtop and is consistent with the lake filling in a few hundred years.

Like much of the pre-climactic eruptive materials, the Wizard Island plat-

Box 7-1
Pillow Lavas

When lava is extruded underwater the surface is quickly cooled, forming a solid pillow-shaped crust while the interior remains fluid. In much the same way that toothpaste is squeezed out of a tube, the still liquid lava inside breaks through the crust to create another pillow. Thus, pillows tend to pile up one atop another, but may break loose to tumble down-slope. These elongate lobes are common in basalts but can form in any composition of lava.

form is composed of andesite. The cinder cone itself consists of small fragments of frothy andesite, while blocky andesite lava flows emanate from its base. Lava flow lobes are readily seen along the western base of the island, and sinuous channels in the drowned portion are depressions between older lava flows. Some of the most interesting hiking in the park is on Wizard Island. The 1^+ km (0.8 mi) trail to the top provides a panoramic view of the interior of the caldera (Plate 15). At the summit, you can explore a well-developed crater some 125 m (400 ft) across and 25 m (90 ft) deep. Following the trail on the west side of the island leads to the emerald-colored pools in its basal lava flows.

Central Platform Volcano – Submerged near the center of Crater Lake is the central platform volcano (Figure 7-4). Its slopes are nearly uniform at angles of 30 to 37 degrees to the horizontal, but less steep "aprons" occur along its north and east base. The edifice of this second-largest volcanic feature in the caldera covers 3.2 km^2 (1.25 mi^2) with a volume of 0.76 km^3 (0.18 mi^3). In addition, lava flows extending from its base into the deeper portions of the basin cover nearly 5.0 km^2 (1.9 mi^2) and have a volume of 0.30 km^3 (<0.1 mi^3). Samples collected during exploration with Deep Rover were andesite in composition, similar to those found on Wizard Island and Merriam Cone.

Earlier speculation, based on reddish oxidized samples, indicated the volcano was formed above water. However, additional (and better) data collected during the 2000 survey, and samples of pillow fragments and angular breccia, suggest a history similar to Wizard Island. Breaks appearing on its slopes indicate the location of shoreline benches and beaches developed during the rising water level of Crater Lake. Below its steep north and east flanks, however, are sinuous lava flows that apparently flowed underwater, down over the shattered older lava. One prominent bench occurs at the same depth on both Wizard Island and the central platform, suggesting that both vents may have been erupting at the same time.

Extensive and prominent lava channels or collapsed lava tubes, (Figure 2-2) have been found on the top of this feature that appear to feed the deeper flows at its base. These features are 30–40 m (100–135 ft) deep and about 100 m (300 ft) wide – one is nearly 4 km (2.5 mi) long. Interpretation of these features suggests that the central platform was constructed as a lava delta similar to the underlying platform of Wizard Island. Toward the end of its active life, individual lava channels or tubes developed that flowed into the lake to form the fan-shaped patterns seen in the basins. One hypothesis is that the central platform was fed by degassed magma from Wizard Island.

Figure 7-8 Merriam Cone *– This view of Crater Lake's floor is looking north with Pumice Point (left) and Cleetwood Cove on the caldera wall. Debris avalanches can clearly be seen along the base of the slopes. The flat areas on either side of Merriam Cone are the Northwest Basin (left) and the East Basin. (USGS image.)*

Merriam Cone – Merriam Cone, possibly the last of the post-caldera andesite volcanoes, lies completely underwater near the north shore of Crater Lake (Figures 7-4 and 7-8). Like so many features in the park, this symmetrical underwater hill was named by Williams for a former director of the Carnegie Institution of Washington. Its summit is 430 m (1,450 ft) above the nearby East Basin (deepest part of the caldera) and less than 150 m (500 ft) below the lake's surface. With a basal area of about 2.4 km^2 (<1.0 mi^2) and a volume of over 0.34 km^3 (<0.1 mi^3), it ranks third among the volcanic features on the caldera floor. Dredged materials from Merriam Cone have a composition similar to the central platform.

Its symmetry and slopes (30°–32°) suggest a cinder cone, but there is no summit crater. Although earlier interpretations indicated that it erupted above water, analysis of the 2000 data suggests otherwise. Images from the remote observation vehicle show jointed blocks or pseudopillows and fracture patterns most likely formed by rapid chilling of lava. In addition, dredge samples lacked surface oxidation that would be expected if they had been exposed to air. Thus, a recent, more favored interpretation indicates that Merriam Cone's volcanic materials were ejected underwater.

Rhyodacite Dome – On the eastern flank of the Wizard Island volcano and overlapping the central platform is the rhyodacite dome (Figure 7-4). This circular feature's summit lies just 30 m (100 ft) below the surface of Crater Lake. With an area of only 0.61 km^2 (0.24 mi^2) and volume of less than 0.08 km^3 (0.02 mi^3), it is the smallest of the post-collapse volcanic features. From observations using the Deep Rover, its surface looks smooth but lightly striated. These smooth upper surfaces and slopes appear to be lava flows, while lower portions have fragmented material that may include pseudopillows. Most of the dome's margins have slopes between 32 and 34 degrees to the horizontal and are composed of rhyodacite talus or breccia, with a composition more silicic than the other post-caldera volcanoes.

An earlier assumption, prior to most of the basin exploration projects, suggested that this feature was a Pelean type dome, similar to the one formed in the Mount St. Helens explosion crater. This appears to be a good interpretation as it may be of the type built by effusion of viscous lava erupted from a centrally located conduit. A core recovered from the central platform contained an ash bed with a composition similar to that of the dome. Sediment containing carbon found beneath this ash bed suggests the dome erupted about 5,000 years ago. This would make it the youngest volcanic feature in Crater Lake National Park. Silica-rich domes are common features associated with ring fractures in larger calderas. However, this one is located too far from the caldera wall to be associated with a potential ring fracture.

Depositional Features

Basins – At Crater Lake the loose debris remaining from the collapse of Mazama rested on slopes along the inside of the caldera. Indeed, much of the existing wall exposed above lake level still displays loose material, and periodically a landslide can be seen. This process would likely have been common for some time immediately following the collapse. Two types of deposits are recognized that result from mass wastage of the caldera walls: turbidity currents and landslide/debris avalanches (Box 7-2).

Beneath Crater Lake

Until the 1960s, little attention had been directed toward the possibility of sedimentation on the caldera floor. The prevailing concept viewed the flat floor topography as resulting from post-caldera volcanic activity, perhaps a lava lake as seen in some other calderas. In addition, with the age of the caldera placed at just a few thousand years, not enough time had elapsed for any significant thickness of sediments to accumulate. However, recent research has shed new light on the three smooth, level basins (Figure 7-3) that occupy more than 40 % of Crater Lake's floor. Earlier speculation assigned these flat regions to lava flows filling the basin following Mazama's collapse. Since then, results from field investigations, including sediment cores and seismic profiles in the 1980s, suggest a much different interpretation. Researchers found that coarse material collected in aprons along the base of the caldera walls. Then, finer sediments were carried farther out into the basins by turbidity currents and deposited in thin, well-sorted layers. These layers consisted of particles ranging in size from sand down to mud (clay), with the proportion of sand decreasing with distance from the caldera wall.

Box 7-2
Turbidity Currents

These features occur when loose materials mix with a fluid, usually on steep slopes below water. The fluid, laden with suspended sediment, moves rapidly downslope under the influence of gravity. It is more dense than the surrounding clear water and forms a mass that hugs the bottom. As it loses energy, due to reduction in the slope, the sediment being carried is spread out over a horizontal surface. When motion ceases, particles settle according to size to form beds. At Crater Lake these features appear to result from the collapse of loose debris slopes along the inside of the caldera.

When a turbidity event occurs, sediment particles mix with water to form a current that is heavier than the water. As a consequence, it flows in a channel-like manner until its energy is used up and then slowly settles out in a layer on the level portion of the lake bottom – in the three basins. Although there is no way to know how often such an event occurred or how many may have deposited sediments on the caldera floor, they undoubtedly occurred frequently, perhaps numbering in the hundreds.

Lying between Merriam Cone and the Wizard Island complex is the Northwest Basin. This depression is thought to have a thickness of sediments less than 50 m (160 ft), mostly from the adjacent caldera walls. Southeast of Merriam Cone is the largest of the basins, the East Basin, where up to about 75 m (250 ft)

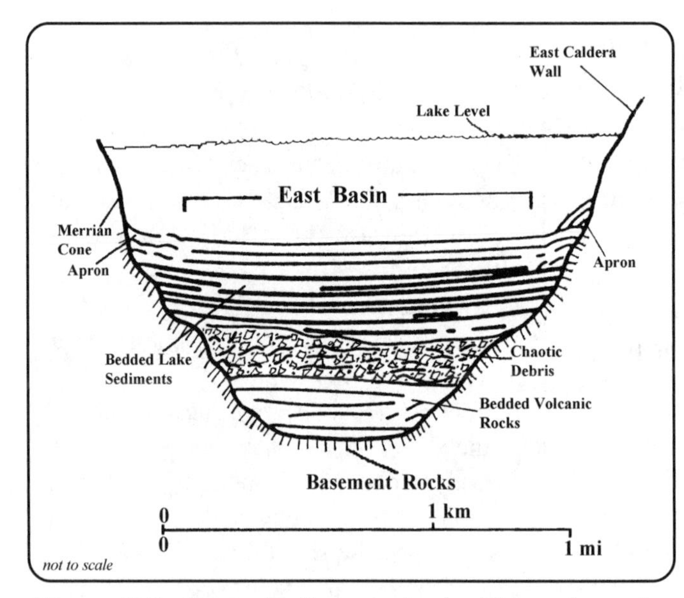

Figure 7-9 Basin Sedimentation *– This schematic section across the East Basin is an interpretation of a seismic profile. It runs E-W from Skell Head to the east slope of Merriam Cone. (After Nelson, 1994.)*

of sediment has been deposited. Finally, the Southwest Basin lies immediately southeast of the Wizard Island complex, with a sediment thickness similar to the Northwest Basin (Figure 7-9).

Along many of the caldera margins, both below and above lake level, another type of topography developed from debris slides: landslides and debris avalanches. These lake floor deposits originating from the caldera walls are characterized by irregular topography and isolated blocks. Many of the scalloped bays around the exposed portion of the caldera mark the location of landslides and debris avalanches. They represent evidence of large mass wastage events in the past like the minor events seen on the walls today. Speculation is that many of these older occurrences were initiated by earthquake activity in the region. These features are best developed in the older portions of Mount Mazama along the south and southeastern caldera wall (Figures 7-4 and 7-10).

The best (and largest) example of a landslide and associated debris avalanche is the Chaski Bay Slide that separates the Southwest Basin from the East Basin. On the caldera wall in this location a large slump-block can be readily observed (Figure 7-10 and Plate 3). Below lake level, remnants of the failed section form large down-dropped blocks up to 200 m (656 ft) in length. Chaski Bay Slide debris covers more than 4.5 km^2 (~1.7 mi^2) of the caldera floor with some 0.22 km^3 (0.05 mi^3) of failed wall rock. Other landslide deposits are found at Danger Bay, Llao Bay, west of Eagle Point, and in the East Basin.

Hot Areas

***Figure 7-10 Debris Avalanches** – This view of the southeast portion of the caldera floor, with Kerr Notch and Sun Notch along the walls reveals a major debris avalanche below Chaski Bay Slide. The East Basin is on the left and the margin of the central platform is in the lower left.* (USGS Image.)

Outside the caldera there are no hot areas to suggest remnant volcanism. Fumaroles, mud pots, steam vents, or other signs that indicate magma still exists below old Mount Mazama have never been observed (Figure 9-8). Still, an 8,000 year old volcanic feature as large as the Crater Lake caldera would be expected to show a significant amount of residual heat. Until the recent surveys (starting in the 1980s), no evidence of leftover volcanic heat had been found. However, this should not be considered unusual as the large accumulation of cold water could easily mask heat emanating from the bottom of the lake. Recent evidence from heat-flow measurements on the lake floor do suggest a source of heat still exists below.

In the 1980s and 1990s, a number of investigations were conducted to check on possible "hot" areas on the caldera floor. This research located several with higher than usual temperatures and other indicators of modern hydrothermal circulation, heat flowing from below (Figure 7-4). Venting of warm water that hosts bacterial mats was found in the southern region near the margin of the Chaski Bay deposits. In the northern part of the caldera, pools of warm, chemical-rich water were located at the tip of the Cleetwood Cove debris deposit. This location also marks the area of the highest measured heat flow in Crater Lake to date. And, during a Deep Rover run in the East Basin adjacent to Skell Head, some spires appear that may be related to thermal vents. All these features occur in places where a ring fracture system would occur. Thus, it seems there is good evidence for thermal activity and the presence of a ring fracture in the Crater Lake caldera. Of course,

***Figure 7-11 Llao's Hallway** – along Dutton Creek in the western portion of the park illustrates the dramatic erosion of Mount Mazama's eruptive materials. Debris seen here is the medial facies of the ring-vent phase that filled valleys with climactic ejecta.*

the most likely interpretation of these findings is residual heat from cooling magma below.

Another interesting phenomenon may also fall into this same category: circular "holes" that have been reported in Fumarole Bay along the southwest side of Wizard Island. These were originally reported as fumaroles formed prior to the lake level reaching its current height. This interpretation has been questioned, and an alternative (and unlikely?) explanation suggests the depressions are tree casts resulting from a lava flow that trapped trees in an upright position. At this time there is no consensus on which of these ideas, or yet another interpretation, is most likely correct.

Other Features

When the central block, an oval mass estimated to be 5 by 6 km (3^+ by 4 mi), subsided along the ring fracture, it left unsupported walls that were nearly vertical. These failed by sliding into the depression along a scalloped pattern, giving the caldera walls their appearance as we see them today. In certain locations on the lake floor, near Eagle Point, Skell Head, and Steel Bay, steep walls may represent the escarpments left by the original ring fracture system.

Mapping and sampling of the caldera walls above the lake provides the sequence of lava flows and pyroclastic deposits from various vents that erupted to build Mount Mazama. Breaks in these eruptive materials are often seen as benches on the caldera walls. Using data from the 2000 survey, similar benches can be recognized below the water on submerged bedrock exposures. In some places, generally less than 150 m (500 ft) below the surface, benches (and rocks) can be correlated with those on the caldera walls above the lake. In a similar way, a few

***Figure 7-12 Godfrey Glen** – The intersection of Annie Creek Canyon with several other valleys illustrates post-caldera errosion that has occurred in the ring-vent phase ash flows.*

dikes have been identified below lake level, with two appearing in the Phantom Cone structure (Figure 4-11). All these dikes, of course, represent activities and rocks emplaced long before Mount Mazama's collapse.

Outside the Caldera

As noted in Chapter 4, running water is an important agent of erosion and is undoubtedly the major cause of erosion outside the caldera since the collapse. Even a casual inspection of the streams flowing down Mazama's slopes today will verify this conclusion. As the climactic ash flows poured down the pre-collapse stream valleys, their original profiles were covered and/or buried with as much as several hundred feet of pyroclastic material. When streams reestablished their valleys in these ash flows, rapid erosion by running water quickly cut down into or through the loose material (Figure 7-11). The result of nearly eight thousand years of erosion can readily be seen along park roads bordering Annie Creek, the Pinnacles, and Castle Creek (Figure 7-12).

Crater Lake, as we see it today, is the result of not only its long volcanic history of mountain building and caldera formation but also the climatic conditions of the last few thousand years. The 15 m (50 ft) of annual snowfall has played a major role in creating the Crater Lake scenery. Following the collapse, high precipitation has both filled the caldera and sculptured the surrounding topography. Is this the conclusion of the Crater Lake story, or will there be other more exciting chapters written by future geological events? Chapter 9 looks at this question. However, before considering the park's geological future, we will examine the physical characteristics of the most pure and deepest body of water in the United States.

Additional Reading

See Bacon, 2002, and Nelson, 1994 in the Appendix references.

Chapter 8
Crater Lake

While the construction and subsequent destruction of Mount Mazama to create the Crater Lake caldera is the real story here, the lake is certainly the most spectacular feature and centerpiece of the park. (Table 8-1 on page 132.) Located inside a volcano, it is the deepest lake in the United States and seventh deepest in the world. Among many of the lake's unusual features is the clarity of the water and its deep-blue color. For those who have never visited the park before, their first sight of its unbelievable location and blue water is breathtaking! People are frequently overheard marveling at the spectacular blue color, just as the members of the Hillman Party did over 150 years ago. If it is not the most beautiful body of water in the world, it certainly ranks high in the competition. Crater Lake is a world-class natural wonder, a national treasure, and Oregon's crown jewel of the Cascades. (See the cover photograph for one view.)

In addition to its color, there are numerous other unusual aspects about the water and nature of Crater Lake. Its source, depth, accumulation, purity, clarity, temperature, and life are all fascinating facets of this unique body of water. All of these are related to or directly caused by the physical characteristics of Crater Lake's water and great depth. For well over a century, much of the research in the park has focused on the nature of the lake and is ongoing even now (Figure 8-1).

Figure 8-1 The "Old Man" *– This Mountain Hemlock log floats with the currents in Crater Lake. It is about 30 feet long and floats vertically like a fishing bobber. Although the "Old Man in the Lake" has been sighted for decades, it is not known how or when it became a part of Crater Lake's lore. It is often seen by visitors on boats for the geological tour to study the inner caldera walls.*

Physical Characteristics

Depth and Source – Crater Lake has been known as the deepest lake in the United States since 1886 (Figure 1-4). Interestingly, that original depth measurement of 1,996 ft (600+ m) was determined using a weight and wire line, resulting in nearly the same value as that obtained by the much more sophisticated survey conducted in the year 2000. The difference in maximum depth (1,996 ft versus 1,953 ft) between the two measurements, separated by over a century, is only about 15 m (~50 ft) or just 2.5 %. Much of the lake, however, is not as deep and averages about 350 m (1,150 ft). A surface elevation of 1,883 m (6,178 ft) above sea level has been set as a benchmark for Crater Lake by the United States Geological Survey and is the value used for all recent research conducted on the lake.

The area that supplies water to Crater Lake is only about 68 km^2 (26 mi^2), and the lake covers nearly 80 % of its drainage basin. Consequently almost all the water reaching the lake falls directly into it as either rain or snow. The balance of the drainage surface that feeds the lake is from the caldera wall that slope inward. While there are no permanent surface streams flowing into or leaving Crater Lake, over fifty "springs" have been identified along the caldera wall representing inflow of groundwater. Most of these are small, but a few are spectacular, and in some places they fall directly into the lake (Figure 8-2). In addition, groundwater is certainly being added below lake level, but the locations and amounts are unknown. Thus, most of the input is

Figure 8-2 Waterfall *– "Springs" along the southern caldera wall mark the locations where groundwater flows into Crater Lake.*

directly from precipitation (averaging about 168 cm or 66 in./year), primarily snowfall (13 m or 44 ft/year on average) that occurs between October and May. A drop of water entering Crater Lake will remain there for about 150 years, its so-called residence time. Another way to think about residence time is the average amount of time required to refill a basin with new water if it were to be empted.

Figure 8-3 Lake-Level Device *This lake-level measuring device was used to check and record Crater Lake's elevation during early lake surveys.*

Filling of Crater Lake – From a geological perspective, the filling of the Crater Lake caldera is one of the most recent events to occur in the park. Water began accumulating in the caldera soon after Mazama's collapse. Originally, water likely came from groundwater draining into the basin from surrounding slopes and hydrothermal springs on the caldera floor. Assuming the climate was similar to our present one, some 57 billion liters (15 billion gallons) of water were added each year. This resulted in a rapid increase in depth, perhaps bringing Crater Lake to about half its present depth in 150 years or so. If this rate persisted, at 1 m/year (3-4 ft/year), the lake would approach three-quarters of its current depth in about 300 years. In 700 years it would have reached an equilibrium between water gained and lost and achieved its present level (Figure 8-3). Alternative models, however, suggest a different story. Some evidence indicates that a drier climate period began about 500 years following the collapse and lasted some 1,000 years. Depending on many factors and the assumptions used, continued filling of Crater Lake may have proceeded in a number of different ways, as illustrated in Figure 8-4. Based on this particular set of assumptions the current lake level of 1,883 m (6,178 ft) was reached about 2,000 years after the climactic eruption, collapse, and formation of the basin.

As Howel Williams clearly stated, there does not seem to be any way to

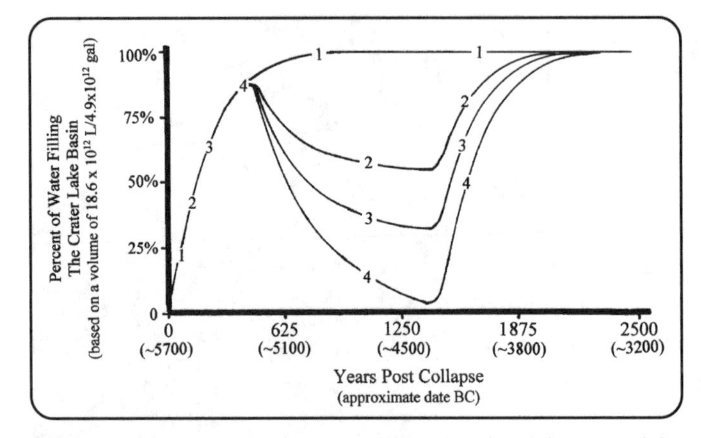

***Figure 8-4 Filling Models** – This graph of four models illustrates how Crater Lake may have filled over time. Model 1 assumes that present-day annual precipitation has been constant over time. Models 2, 3, and 4 assume, for a 1,000-year altithermal period that prevailed between 500 and 1,500 years post collapse, a reduced precipitation rate that is 0.7, 0.6, and 0.5 times that of the present, and an evaporation rate that is 10%, 15%, and 20% greater than the present rate, respectively. (After Hoffman, 1999.)*

determine what time span was required for the caldera to fill to its present level. There are simply too many unknown factors to allow an accurate determination. On the other hand, there is no strong evidence, such as higher beach lines, to indicate that the lake level was significantly higher than now. However, Williams noted that a member of the National Park Service staff "... discovered diatomaceous earth on Wizard Island approximately 50 ft [~15 m] above the surface of the lake," suggesting the possibility of a higher level at some time in the past. Diatomaceous earth is largely composed of the hard parts of diatoms, a freshwater algae, that would have accumulated along the lake's shoreline.

Early investigators set up a number of experiments designed to establish the relationship between how water is added to and removed from Crater Lake. These were primarily concerned with evaporation from the lake's surface and consid-

Box 8-1
Crater Lake Water Balance

The amount of water in Crater Lake is determined by measuring the precipitation entering the lake and water loss through evaporation and leakage. Using data from precipitation measurements, it is estimated that the average amount of water added to the lake, over the long term, would raise its level about 169 cm/year. Calculations indicate that evaporation accounts for 85 cm/year (~50%) and seepage is 84 cm/year (~50%). While this might seem like an easy calculation, comparing the amount of incoming water with that leaving, it is more complicated. (Based on USGS Open-File Report 92-505.)

eration of how water might escape by percolation through porous zones. Results of these measurements suggest about two-thirds to three-fourths of the water is lost by surface evaporation and the balance by seepage through porous rock layers. See Box 8-1 for another interpretation. Crater Lake's hydrological features have been observed and recorded over the past 120 years. In the past 100 years the surface elevation of Crater Lake has fluctuated within a range of only about 5 m (16 ft). On a yearly basis the level has fluctuated only about a meter or so (3–4 ft).

Figure 8-5 Annie Springs *– This large spring along the road near Mazama Campground is typical of many others found on the flanks of Mount Mazama. It is a major source of water for the park.*

Numerous springs (Figure 8-5) on the outer flanks of Mount Mazama may represent escaping lake water. There has been speculation that seepage from Crater Lake contributes to the initial flow of the Rogue River at its source, Boundary Springs, in the northwestern corner of the park. Another likely possibility is the Wood River, whose source is the Wood River Springs. Recent chemical evidence from this large spring near Fort Klamath, about 20 km (12 mi) southeast of the lake, indicates the water may be from Crater Lake. Thus, the current lake level and its relatively constant surface elevation appear to have developed primarily from precipitation over its drainage area with loss due to evaporation and seepage. Together, these keep the lake's volume nearly constant, and based on lake-

Figure 8-6 Lake-Level Station *– This permanent lake-level gauging device is located at the bottom of the Cleetwood Trail on the north side of the caldera. This trail is the only access to the lake and boat docks.*

level measurements obtained for over a hundred years, the lake appears to have reached an equilibrium (Figure 8-6).

While models can suggest how all this might have happened over the last several thousand years, they are largely a matter of speculation using a few observations, certain facts, and estimates of past climatic conditions. Certain facts have been known for years and include details of water temperature, depth, volume, surface area, average precipitation, and the topographic configuration of both the lake floor and land surface. Such models are based on calculations using precipitation records and estimates of evaporation from the lake surface, along with assumed percolation into porous zones along the sides and bottom of the basin. A porous zone under the Palisades, below lake level, has been postulated as a major location where water is lost.

Purity and Clarity – Almost as unusual as its color is the purity of Crater Lake water. As one researcher noted, it is "intermediate between dust-free and distilled water." Indeed, 80% of all its water is distilled water in the form of snow or rain falling directly into the lake. An additional 16% is snowfall or rain that rapidly reaches the lake from the surrounding basin walls with little opportunity to assimilate either sediment or dissolved matter. Chemical analysis reports show most dissolved elements and compounds at a few parts per million, extremely low for surface water. However, Crater Lake water is relatively high in dissolved solids, both organic and inorganic, when compared to other lakes in the Cascade Range. More important, Crater Lake's spectacular beauty is, perhaps, more closely related to the near absence of particulate matter suspended in the water. One interesting aspect of this is the transparency of Crater Lake's water and the depth to which light will penetrate. Measurements over the years have frequently indicated visibility to depths well over 40 m (130 ft) (Figure 8-7). One report, from a dive with the Deep Rover, indicates visible light can even be seen on the bottom (Buktenica, M., 1996).

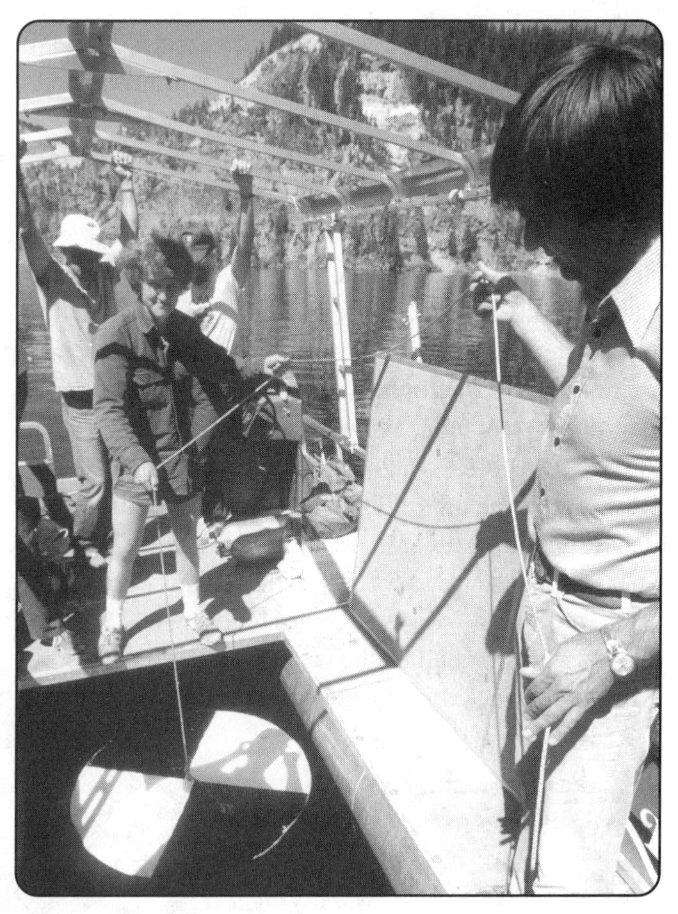

Figure 8-7 Secchi Disk *– This circular device is 1 m (3 ft) in diameter and painted black and white to improve its visibility. It is used to estimate the distance it is possible to see into the water, a measure of lake water clarity. (Image by D. Larson.)*

Color – Certainly, most visitors would agree that the single most unusual feature of Crater Lake is the water's color. Although its color may vary over a wide range of hues, the deep indigo is the most dramatic. This is related to a number of factors, including water depth, transparency, purity, suspended matter, and steepness of the caldera walls. The most important cause, however, is associated with the nature of water molecules and their ability to absorb and scatter light. When light (solar radiation) enters the lake, it penetrates through the water column by a process known as downwelling. As the light continues deeper, it is gradually absorbed and scattered. Because of the lake's extraordinary clarity, due to the scarcity or near absence of suspended particulate matter, light is able to penetrate to great depths. Roughly 1 % of the original light entering at the surface still remains at depths of around 100 m (~300⁺ ft). In Crater Lake, the water itself is the main absorber of light, while the scattering of light is done almost entirely by water molecules, a process called molecular scattering. Water molecules tend to scatter the short wavelengths of light, which comprise the blue segment of the visible light spectrum. A portion of this scattered blue light is directed back toward the lake surface (backscattered light), which visitors see as the deep blue color. As a result, most of the colors in Crater Lake are lost, with only the blue portion of the original sunlight entering the water being returned back toward the surface.

Temperature – As might be expected, based on the source of the lake's water and its elevation, Crater Lake is cold (Figure 8-8). A typical summer temperature profile is shown in Figure 8-9a. A maximum surface temperature of about 18°C (65°F) is reached in late summer at the surface, but the average is only 13°C (55°F). Then, the temperature decreases with depth down to about 100 m (~325 ft) where a temperature of less than 4°C (39°F) continues all the way to the bottom (Figure 8-9b). During winter months, surface heat is lost to cold air until the surface is at a near freezing temperature. However, rarely has the surface of Crater Lake completely frozen

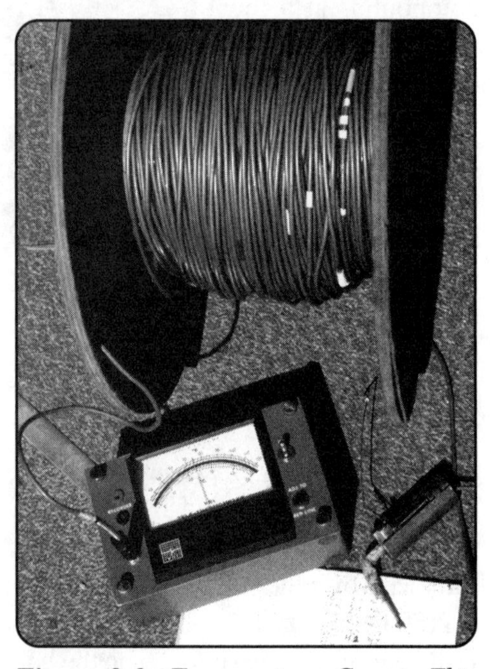

***Figure 8-8 Temperature Gear** – This sensor, meter, and connecting cable were used to measure temperatures in Crater Lake during the 1970s and 1980s.*

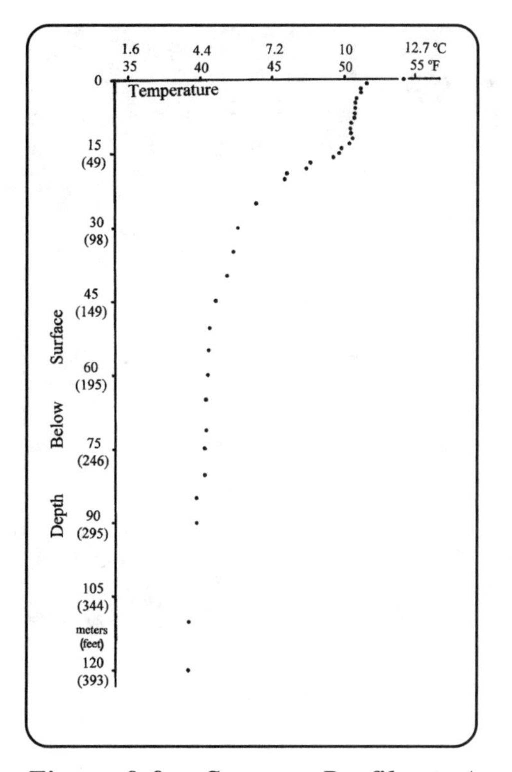

Figure 8-9a Summer Profile *– A typical temperature profile for near-surface water at Crater Lake. (Plotted from measurements taken in August 1982.)*

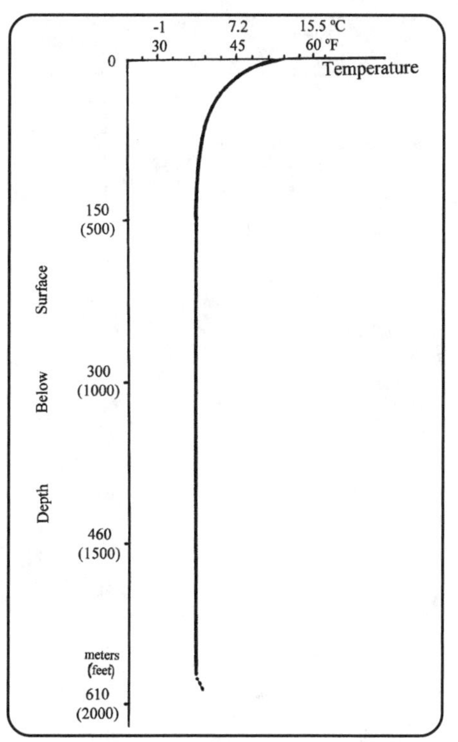

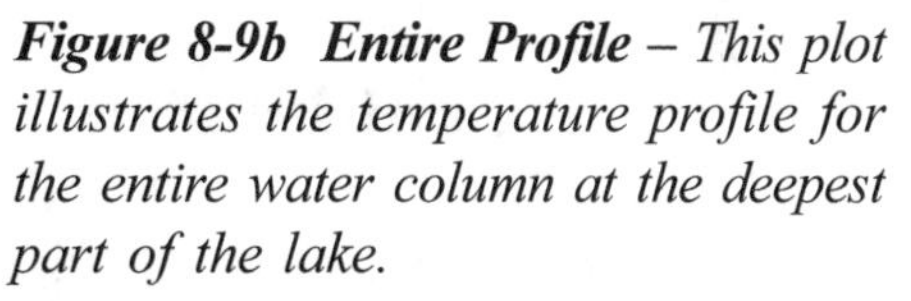
Figure 8-9b Entire Profile *– This plot illustrates the temperature profile for the entire water column at the deepest part of the lake.*

over. This unexpected condition results from several factors, including the large volume of water, due to the lake's great depth, and surface currents driven by the strong winter winds. During the summer the amount of heat absorbed (its heat income or heat storage capacity) is relatively high – determined to be 36,000 gram calories/cm^2 in 1968–69. Summer heat income is defined as the maximum amount of heat necessary to raise the temperature of the lake from 4° C (39°F) to its highest heat storage content. Because of this high heat storage capacity, the entire lake surface seldom cools down enough during the winter to form ice, that is, 0°C (32°F). Strong wind action inside the caldera during winter also contributes to maintaining open-water conditions but is less important than high summer heat income. Thus, over the course of a typical winter season, surface water seldom reaches freezing temperatures long enough to form ice over the entire surface (Figure 8-10).

Figure 8-10 Frozen Lake *– This 1949 view of a completely frozen Crater Lake is one of the few times the entire surface has been covered with ice.* *(NPS image.)*

Biological Characteristics

Crater Lake is known as an oligotrophic (meaning low productivity) lake in terms of the living organisms found in it. There are several factors that contribute to this condition. First, because the lake is located at an elevation of nearly 2,000 m (6,200 ft), the growing season, when light and temperatures are favorable for the growth of aquatic plants, is relatively short, roughly mid-June to late August. Second, the lake's great depth, steep-walled basin, and the nature of its shoreline also contribute to Crater Lake's relatively low organic production. Near-shore waters are too deep for light to reach the bottom, except for some of the shallow-water areas around Wizard Island. This condition precludes or greatly restricts the development of near-shore beds of aquatic plants. In addition, much of the shoreline is bare and exposed, with few protected indentations, such as coves or bays that favor the production of organic matter. Third, nitrogen appears to be the limiting nutrient needed for greater production of organisms in Crater Lake. Other nutrients, such as phosphorus and silica, are more abundant and adequate to allow more organic production than actually occurs. Nutrients are recycled by bac-

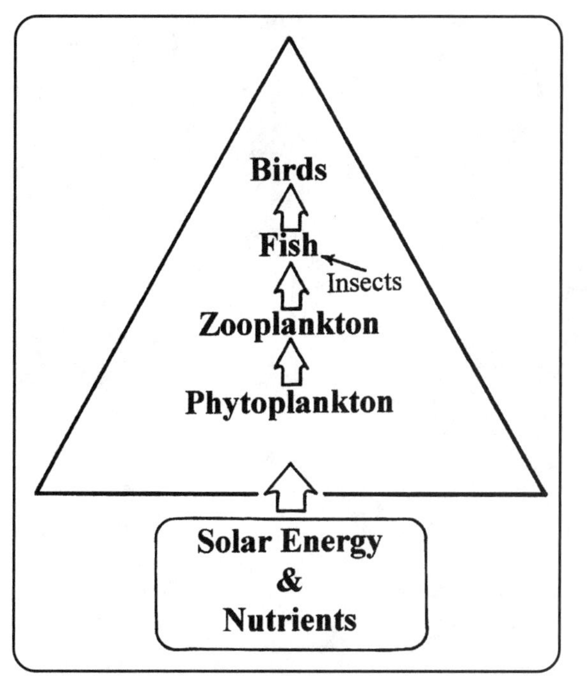

Figure 8-11 Food Web – This simplified sketch illustrates the typical food web in high mountain lakes such as Crater Lake.

teria and used by phytoplankton which in turn feeds zooplankton, the next step on the food web (Figure 8-11).

Near-shore portions of the lake around Wizard Island do support beds of aquatic plants. These areas are also habitats for substantial populations of aquatic insects and other invertebrates. Thus fish are present, including large rainbow trout that feed on the aquatic invertebrates. Zooplankton are also abundant in these rare shallow zones, supplying food for kokanee salmon. Research indicates that these fish "attain comparable size and growth rates of fish of the same species in other Northwest oligotrophic lakes." A few larger invertebrates and insects make up most of the larger forms of life in Crater Lake, with fish representing the top of the food pyramid. Although there appears to be little organic production in Crater Lake, it is not a sterile environment in which life can barely exist. Thus, the suggestion that its waters are similar to distilled water appear to be something of an exaggeration.

The commonly asked visitor question "Are there fish in Crater Lake?" has already been answered. Originally it was thought there were no fish in the lake. However, following its discovery, early visitors began stocking the lake (first recorded in 1888) with rainbow, brown, and cutthroat trout as well as steelhead, coho, and kokanee salmon, although the kokanee was supposedly an accidental introduction. This practice was continued until 1941 and may have had a detrimental effect on the original populations of other organisms. Today only rainbow trout and kokanee salmon remain from the initial stock of fish. Fishing is allowed, even encouraged, to assist in removing the fish, as there is some interest in restoring Crater Lake to its original condition (Figure 8-12).

Recent Research

Figure 8-12 Nice Fish *– Large rainbow trout like this one have been caught at Cleetwood Cove and from Wizard Island.*

In 1982 Congress mandated a ten-year study to monitor limnological conditions in Crater Lake. Since then, the National Park Service has kept it going and added a research program as well. This has provided the opportunity and resources to study many aspects of Crater Lake, ranging from its chemistry to biology over the past quarter century. The initial monitoring efforts used surface boats and traditional methods such as temperature and clarity measurements, chemistry of the water at various depths, and continued sampling of organisms. Gradually, however, the research program began using more sophisticated techniques. Rock samples collected from the caldera floor proved valuable in understanding the post-collapse geology. Probes lowered into the sediments of the floor collected samples and measured temperatures. Results from these lead to a much better interpretation of the caldera floor geology and possibility of hydrothermal springs and/or fumaroles below the lake's surface (Figure 8-13). The monitoring program is supported by the National Park Service, while most of the research is conducted by the United States Geological Survey and other institutions.

Surface vessels (Figure 8-14), a remotely operated vehicle (ROV), and a manned submersible vehicle (Deep Rover, Figure 7-2) have been used to examine deep portions of the caldera floor. The ROV was employed in 1987 to produce video images to depths down to 500 m (1,600 ft). It provided the first good images of the caldera floor at depth and assisted with the geological interpretation of what happened following Mazama's collapse. A total of 47 dives were made using Deep Rover during the 1988 and 1989 seasons to study Crater Lake's floor. Several were suggested by Bacon to investigate specific portions of the caldera floor. Discoveries from these dives have verified a number of ideas developed from surface measurements and added details of the caldera floor.

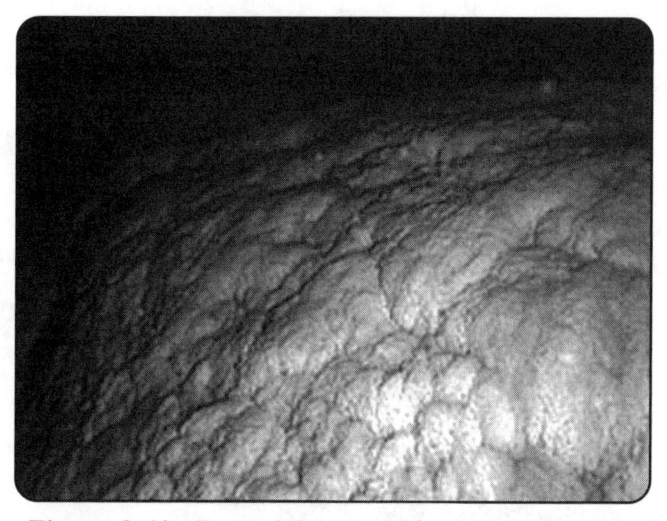

Figure 8-13 Bacterial Mat *– These communities of bacteria appear to be concentrated above the ring-fracture system on the lake floor. Elevated temperatures, as much as 15°C (27°F) higher than surrounding water, and much enriched fluids indicate hydrothermal water is being vented on the floor of Crater Lake. Small pools with enriched water may also mark the location of hydrothermal vents, but temperatures are much lower, only 1 or 2°C (~2-4°F) above the surrounding water. Both of these indicate that hot volcanic material is still present below the caldera, providing the energy for fluids moving upward and venting into the lake.* (*NPS image.*)

Direct observations, video images, physical measurements, and samples collected (both rocks and biological specimens) were accomplished on the Deep Rover dives. Nearly half of the dives were dedicated to collecting geological information, especially locating and investigating hydrothermal input to Crater Lake. Much of the geological information derived from these dives is incorporated in Chapter 7. In addition, the dives dedicated to the biology of the lake floor discovered some interesting phenomena. Dense beds of mosses were found, primarily in the shallower zones, while a number of different species of algae were discovered to be common in the deeper parts of the lake. Various forms of diatoms, a form of microscopic algae that secrete a siliceous (silica) shell, were also present on rock surfaces and associated with other algae.

Figure 8-14 Research Vessel *– This pontoon boat is typical of those used to conduct studies on Crater Lake since the late 1970s. They provide a stable platform, to lower various instruments into the water.* *(Image by D. Larson.)*

Additional Reading

Buktenica, M., 1996, Why enter a sleeping volcano in a submarine?, *Crater Lake Nature Notes*, V. XXVII. Relates working in Deep Rover during a research dive in Crater Lake.

Hoffman, F. O., 1999, The Filling of Crater Lake, *Crater Lake Nature Notes*, V. XXX. A detailed discussion of possible models that describe the filling of Crater Lake following the formation of the basin.

Mark, S. M., and Gnass, J., 1996, *Crater Lake: The Continuing Story (in pictures)*, KC Publications, Inc., ISBN 0-88714-109-9, 48 pages. A nice collection of photographs and commentary of Crater Lake National Park.

Table 8-1
Crater Lake Facts

	Metric	English
Diameter (at rim)		
E - W	9.76 km	6.1 mi
N - S	7.52 km	4.7 mi
Surface Area	53.4 km^2	20.42 mi^2
Depth		
	greatest: 592 m	1,943 ft
	average: 350 m	1,148 ft
Volume	18.6 trillion L*	4.9 trillion gal*
Surface Elevation	1,882 m	6,173 ft
Water Balance		
gain	128 billion L	34 billion gal
loss by:		
evaporation	60 billion L	16 billion gal
seepage	68 billion L	18 billion gal
Temperature		
	surface 0° to 19°C	32° – 66°F
	bottom ~4°C	~38°F
Transparency	up to 44 m	up to 144 ft

* L = liters, and gal = gallons.

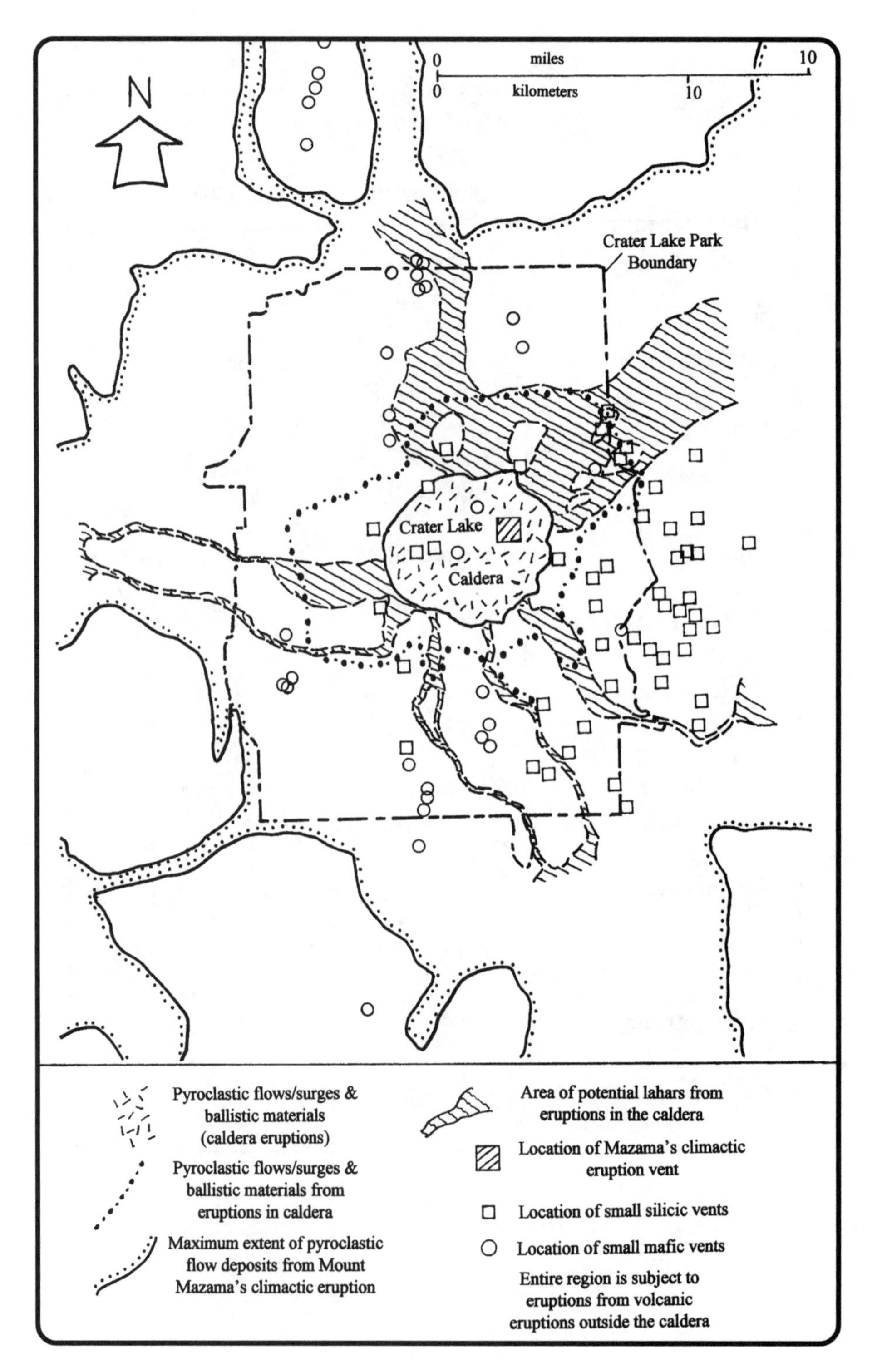
N
0
miles
10
0
kilometers
10
Crater Lake Park Boundary
Crater Lake
Caldera
Pyroclastic flows/surges & ballistic materials (caldera eruptions)
Pyroclastic flows/surges & ballistic materials from eruptions in caldera
Maximum extent of pyroclastic flow deposits from Mount Mazama's climactic eruption
Area of potential lahars from eruptions in the caldera
Location of Mazama's climactic eruption vent
Location of small silicic vents
Location of small mafic vents
Entire region is subject to eruptions from volcanic eruptions outside the caldera

Chapter 9
Crater Lake Tomorrow

As any visitor can readily determine, both Crater Lake and Crater Lake National Park exist because of geological events that have transpired over the last several thousand years in south central Oregon. Like many other national parks, Crater Lake was established to protect and preserve the geological setting and other natural aspects of the area. This commitment by the people of the United States through Congress ensures that future visitors will find the park, and its geology, in a natural state just as it is today. Does this mean that nothing will change? On the contrary, natural systems are dynamic and include geological changes, usually slow as humans reckon time, yet they continue to operate at Crater Lake.

In general, volcanic activity along the Cascades has not been pronounced over the last two hundred years, the period since the region has become densely populated (Table 9-1). Of course, one glaring exception to this observation is Mount St. Helens' eruptions starting in 1980 and its continued intermittent activity.

From a human perspective, this quiet period may be considered typical. However, episodes of much greater volcanism are every bit as typical for these volcanoes. Will Crater Lake share in any renewed activity along this volcanic mountain chain? Geologists find questions like this difficult to answer. Predicting future geological events, especially volcanic activity and earthquakes, has become an area of systematic study. The United States Geological Survey, through its

Figure 9-1 Geological Hazards Map *– This map illustrates some of the possible geological events that could occur in the future in and around Crater Lake National Park.* (Modified from Plate 1 of Volcano and Earthquake Hazards in the Crater Lake Region, Oregon, *[USGS Open-File Report 97-487.]*)

Volcanic Hazards Team, has studied potential geological events along the Cascade Range, including Crater Lake. Mount Mazama's long history of volcanism strongly suggests that this center of volcanism will be active again. This final chapter outlines future geological activities and their likelihood for the Crater Lake region. Figure 9-1 illustrates the areas likely to be affected by various future geological events, and Table 9-2 lists estimates of the probabilities of these events (page 145).

Table 9-1
Cascade Volcanoes Observed in Eruption

Volcano	Eruption(s)
Mount Baker (WA)	Several between 1792 and 1880
Mount Hood (OR)	1866 (Crater Rock enplaced ~200 years ago)
Glacier Peak (WA)	Late 1700s and early 1800s
Lassen Peak (CA)	Between 1914 and 1921
Mount Rainier (WA)	Between 1820 and 1854
Mount Shasta (CA)	1787 and mid-1850s
South Sister (OR)	1853 (?)
Mount St. Helens (WA)	Early 1800s, 1820s to 1870, 1980 and continuing

Other Cascade volcanic areas have experienced earthquake activity that may have been related to movement of magma in the subsurface. (Source: Harris, 1988.)

Volcanism

Renewed volcanism is the most probable geological event at Crater Lake National Park. It might seem reasonable to use the history of Mount Mazama or other Cascade volcanoes as the basis for predicting future activity at Crater Lake. However, the climactic event that produced the Crater Lake caldera resulted in profound changes that will likely affect future volcanism. The remaining magma system has been much modified and is unlikely to produce large eruptions in the future similar to those that have occurred at Cascade volcanoes. There are two scenarios for future volcanism at Crater Lake: (1) within the caldera, the most likely location, and (2) in the region surrounding the caldera. Of course, both of these could occur.

In the Caldera – The last volcanic event in the park was about 5,000 years ago on the floor of the caldera (Chapter 7). This suggests any new or renewed activity would probably develop there. Based on the final eruptions in the park, the western portion of the caldera is the most probable location for future eruptions. Except for the caldera walls, these would develop beneath the water. Such volcanic events are termed hydromagmatic, the violent mixing of lake water with hot magma. Such events have been rare in Mazama's history, so to understand the behavior of such an eruption at Crater Lake we must look to examples elsewhere. The nature of hydromagmatic events depends on many factors, including type of magma (especially its temperature), rate of production, amount of gas involved, and water depth. Steam may be generated when hot fluid rock encounters water. Under favorable conditions, steam will expand rapidly, resulting in an explosion (Figure 9-2). The most violent occurrences are in shallow water with rapid extrusion of magma that fragments easily. In deeper water, such as Crater Lake, pressure of the overlying water column tends to inhibit this process, and violent behavior is less likely.

Should an eruption within the caldera occur that is large enough to produce an explosive event, due to generated steam, fragmental materials could surmount

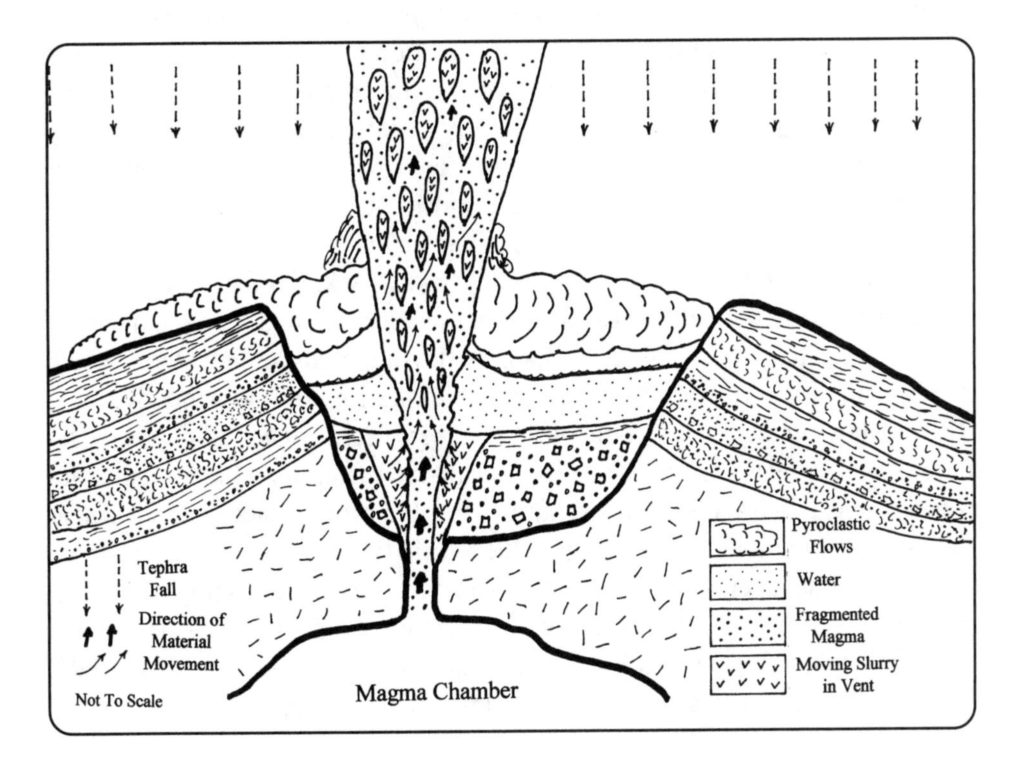

Figure 9-2 Hydromagmatic Eruption *– This hypothetical sketch illustrates what might result if an eruption were to occur on the floor of Crater Lake. (Based on a diagram by White and Houghton in the* Encyclopedia of Volcanoes, 2000.*).*

the caldera walls, producing pyroclastic flows on Mazama's flanks. In a typical case, a plume of ash and debris might extend high into the atmosphere from such an eruption. This would probably not be as large as the climactic column, but similar in behavior. Materials falling out of such a cloud would likely move down existing valleys in much the same manner as the climactic flows (Figure 6-8). The most serious form would be a pyroclastic surge: a mixture of air, volcanic gas, steam, and rock debris that flows along the surface at high speeds, perhaps several hundreds of meters per second (hundreds of miles/hour). Depending on a number of factors, these flows might travel only a short distance or could move many kilometers (miles) downslope.

A less serious scenario is an eruption that does not create the massive explosion described above. In this case, ballistic blocks from a hydromagmatic event traveling at high speeds, hundreds of meters per second (hundreds of miles/hour), could be ejected during the explosion. They could easily overtop the caldera walls and land several kilometers (miles) away if the eruption occurred near the lakeshore.

Still another possible hazard is the generation of waves on the lake. Large waves, several meters (feet) high, could be produced during an eruption in the caldera. While these would not likely be large enough to escape the lake, they could initiate mass wastage activity along the caldera walls. Unstable portions of the confining walls could fail, causing landslides and debris flows like those that occurred soon after Mazama's collapse (Chapter 7).

Should an eruption occur with a heavy snow pack, flooding and/or lahars could result from the large amounts of meltwater produced. Lahars, the most spectacular of these possibilities, are rapidly flowing mixtures of water and rock material (Figure 9-3). They can range from thick slurries to simply muddy water. Rapidly moving meltwater would pick up loose volcanic materials deposited by the climactic events – both rock debris and air-fall materials. This form of mass wastage may reach speeds approaching 20 m/s (45 mph) in steep channels near their source, but move much slower as the terrain flattens out. Like any moving water phenomena, they are confined to existing valleys. Depending on its size, primarily determined by the amount of meltwater involved, a lahar may eventually flow out onto the flats beyond Mazama's flanks. Lahars could also be generated if lake water is ejected by an explosive event within the caldera.

Regional Volcanism – When the Cascade Range is mentioned, we commonly think of the large composite cones like Mount Rainier, Mount Shasta, or Mount St. Helens before its 1980 eruption. Mount Mazama was also in this cat-

Figure 9-3 Mount St. Helens Lahar *– On March 19, 1982, an explosive eruption melted snow, producing this dark-colored lahar. Part of the flow reached Spirit Lake (lower left), but most followed the Toutle River valley and eventually reached the Cowlitz River about 80 km (50 mi) downstream.* (USGS image by T. J. Casadevall.)

egory prior to the climactic event. Much less prominent, but much more abundant, are cinder cones, lava domes, and other small volcanic features found along the Cascades. In geological terms, these had a much shorter life span, usually involving a single eruptive episode, than the massive composite volcanoes. However, due to the continued active nature of the Cascade region and the number of volcanoes, new eruptions of these features should be expected. Thus, they are a potential future hazard – in the Crater Lake area as well as up and down the entire Cascade Range (Figure 9-1).

These small volcanoes and volcanic features are primarily basaltic in composition and represent the "background" volcanism along the Cascade Range. Hazards from any new regional vent(s) would likely include lava flows and tephra falls resulting from possible explosive behavior of an eruption. In the Crater Lake area, taken to be approximately 11 km (7 mi) north and south of the caldera and about 30 km (20 mi) wide, there are over sixty such features. These vary in age ranging from a few thousand years old to over a million. Estimating future events for such features requires knowledge of frequency, regularity, and distribution of their eruptions. Based on a number of factors and assumptions, the annual chance of a new eruption is about 1 in 10,000. This estimate, however, must be seen as only an approximation.

Other potential geological events may occur at Crater Lake, but all have a low probability. They are (1) a large pyroclastic eruption, similar to Mazama's climactic eruption 7,700 years ago, (2) the sudden release of lethal carbon dioxide gas (CO_2) from the lake, and (3) the catastrophic draining of the lake. In contrast, however, the possibility of future earthquakes is much more likely.

Earthquakes

Seismic activity along the Cascades results from three sources: (1) energy released when local near surface faults move, (2) movement along the plates that form the deeper subduction zone below the western margin of North America (Figures 3-2 and 3-5), and (3) movement of magma below the surface (volcanic earthquakes). In general, earthquakes generated along local faults by release of energy are smaller than those produced when subduction zone plates shift. The rate of seismicity along the Cascade Range shows variation, with the portion south of Mount Hood (including the Crater Lake area) being quiet compared to other sections. Recorded earthquakes in the park and its immediate vicinity are rare, and only a few historical events have been recorded. This is probably due to a lack of recording devices and the relatively small magnitudes of the quakes. Recently, with more and better seismographs, additional earthquakes have been noted, especially south of the park in the vicinity of Klamath Falls and within the Klamath Basin (Figure 9-4).

***Figure 9-4 Seismic Station** – Remote seismographs like this one are located on volcanoes along the Cascade Range.*

Crater Lake lies immediately north of the Klamath Basin, a known fault-bounded basin (a graben, Figure 3-7). The west side of Klamath Lake parallels the active West Klamath Lake fault zone and two mapped faults (Annie Spring and Red Cone Spring) in the park that appear to be extensions of this trend (Figure 9-5). Geological field evidence indicates that movement has occurred

***Figure 9-5 Faults and Earthquakes** – This map illustrates the location of faults and recent earthquakes in the Crater Lake region.* (Modified from Figure 1 of Volcano and Earthquke Hazards in the Crater Lake Region, Oregon, *[USGS Open-File Report 97-487]).*

--->

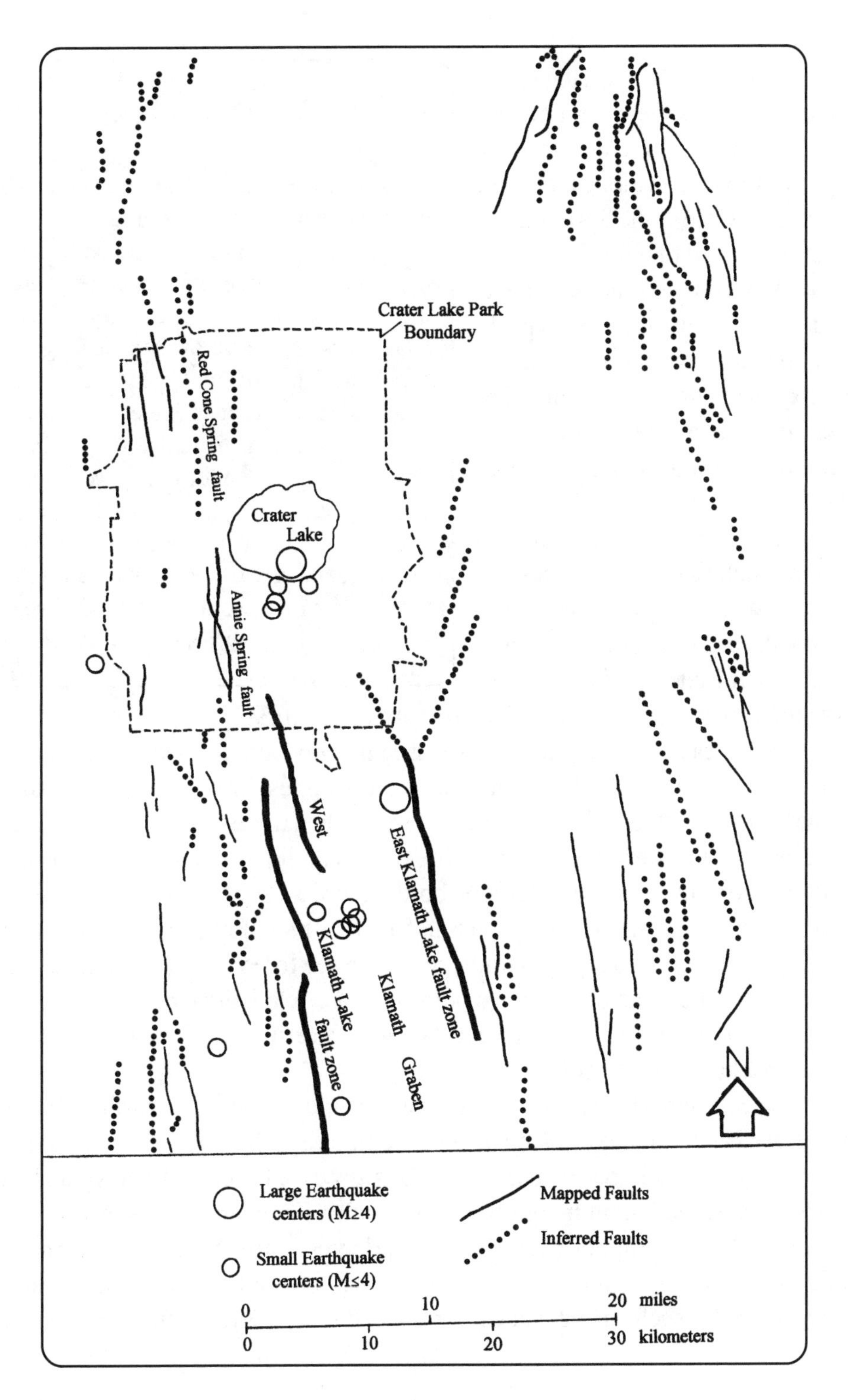
Crater Lake Park
Boundary
Red Cone Spring fault
Crater
Lake
Annie Spring fault
West
Klamath Lake
fault zone
East Klamath Lake fault zone
Klamath
Graben
N
Large Earthquake
centers (M≥4)
Small Earthquake
centers (M≤4)
Mapped Faults
Inferred Faults
0
10
20 miles
0
10
20
30 kilometers

along both these faults in the recent past, thought to be within the last 50,000 years or so. On average, it appears movement along the Annie Spring and Red Cone Spring faults is about a third of a millimeter per year. While this does not sound like much, larger displacements probably occur infrequently to release stored energy as an earthquake. The period of recurrence of motion, and resulting earthquakes, along these faults is unknown. Thus it's not possible to estimate the probability of events such as these happening at Crater Lake. Using the characteristics of the existing faults, the maximum magnitude of a future earthquake in the park has been estimated at about 7 (Box 9-1).

Box 9-1
Earthquake Magnitudes

To compare earthquakes, the amount of energy released at the source must be determined. This is done by measuring the largest seismic wave generated during the earthquake. Using standardized techniques, a value known as the magnitude is assigned to each event. This system, called the Richter Scale, runs from 1 to 10. Each higher number represents ten times the amount of energy released. Thus a magnitude 2 quake would release ten times the energy as one with a magnitude 1, and an earthquake with a Richter number of 6 would have 100 times the energy of one with a magnitude of 4.

The release of energy along the Cascadia Subduction Zone is much larger than that of local faults. Geological evidence suggests very large earthquakes have been generated in the not too distant past when these tectonic plates shifted – perhaps with magnitudes near 9. One likely candidate occurred in 1700 and another along the California-Oregon border more recently (1823). However, as with the local faults, sufficient data is not available to allow prediction of future earthquakes produced along the subduction zone. Moreover, an earthquake produced along the subduction zone would occur at a relatively large distance, perhaps at 100 km (65 mi), so the effect at Crater Lake might be less than a smaller event that happened close by.

When magma beneath a volcano begins to move toward the surface, energy is released as earthquakes. Such quakes are similar to those produced by movement along local faults but can be recognized as associated with a volcano's plumbing system. In general, their magnitudes are smaller. For example, just prior to the Mount St. Helens eruption in early 1980, the largest earthquake had a magnitude of about 5. This occurrence at a Cascade volcano would be a reasonable maximum value for such an event at Crater Lake. It is also noteworthy to remember that a volcanically generated quake would likely occur below the caldera, resulting in a significant amount of shaking.

Other Geological Events

Mazama's climactic event was probably the largest eruption in the Cascades in the past half million years. As described in Chapter 6, it was the culmination of a buildup of silica-rich magma in the mountain's magma chamber over a long period of time. It has been estimated that rhyodacite accumulated at an average rate of about 2 km^3/1,000 years. Current thinking, based on geological evidence, indicates that most of the gas-charged rhyodacite magma was ejected, or cooled and crystallized, following the climactic eruption. However, during the relatively short time since that event, there has not been sufficient time for another large batch to accumulate. Thus the probability of a major silicic pyroclastic eruption is judged to be low, and anything like a caldera-forming event is thought to be negligible.

In some lakes located in volcanic craters or calderas, CO_2 generated from escaping magmatic gases accumulates near the bottom of the lake as dissolved gas. If the lake should suddenly "overturn" (the water at the bottom moves to the surface), the CO_2 is released to the atmosphere. Since it is heavier than air, cold CO_2 tends to flow into low areas and "pond" much like water would. The gas thus displaces air in these low topographic pockets. The danger occurs because it is colorless and not readily detected. Animals, including people, venturing into such locations cannot breathe and are quickly asphyxiated. Although Crater Lake has a source of CO_2 from its underlying magma, the possibility of a buildup of large amounts of this gas is unlikely. Recent lake research suggests that the deeper water regularly mixes with the upper 200 m (~ 650 ft) over a period of between 2.5 and 3.5 years. Consequently, any dissolved CO_2 would be released before a significant amount could accumulate.

The volume of water in Crater Lake is about 17 km^3 (4 mi^3). Should this amount, or a significant portion of it, be released quickly due to failure along a portion of the caldera wall or some other event, the resulting floods and lahars would be catastrophic. However, the amount of rock that would have to be removed is much too large for any logical geological mechanism to accomplish. Another possibility is a major change in climatic conditions that might cause the lake level to rise and overflow the rim at a low point. This would certainly require a significant period of time. Even so, should this happen at some future time, subsequent erosion of a newly established drainage channel would have to re-

***Figure 9-6 Newberry Crater** – Two lakes have formed on the floor of this caldera and drain westward through a low portion of the bounding wall.*

move rock rapidly before a catastrophic flooding event could occur. Since the lowest points along the rim are about 165 m (~500 ft) above the lake, neither of these scenarios seems likely.

Crater Lake's Fate

What is the future of the lake itself? Geologically speaking, the lake will certainly experience significant changes. Lakes are ephemeral features, and few endure for more than a few centuries or hundreds of centuries. Relatively short-term changes will likely be tied to modifications in the climate, especially precipitation and temperatures. Currently the lake appears to be in equilibrium, with evaporation and leakage just offsetting precipitation. However, should a climate change increase or decrease precipitation, the level of Crater Lake would likely respond by rising or lowering, respectively. While specifics of any climate fluctuations are difficult or impossible to predict, they are almost certain to occur in the next few hundreds or thousands of years.

Finally, given enough time, Crater Lake will certainly be drained. If some type of volcanic or other more dramatic event does not occur, erosion will even-

Figure 9-7 Active Area *– This overview of an active hydrothermal system is in the Bumpus Hell area at Lassen Volcanic National Park.*

tually breach one of the walls, permitting water to flow out. While this could happen catastrophically, it would more likely occur slowly, gradually lowering the level of the lake until little remains of what we see today. A somewhat analogous situation can be seen only a short distance to the northeast of Crater Lake at the Newberry Crater caldera (Figure 9-6). At anything like current erosion rates, draining of the lake by such ordinary processes lies far in the future.

Direct Evidence of Volcanism

As noted in Chapter 7, the only known hot areas in the park occur on the caldera floor masked by the overlying cold lake water. However, a survey of other areas in the Cascades finds many locations with hot springs, boiling mud pots, fumaroles, and other direct indications of volcanic heat beneath the surface. Excellent and easily accessible, examples of such features are at Lassen Volcanic National Park in northern California (Figures 9-7 and 9-8). These areas clearly represent heat energy from near-surface magma bodies. Recent drilling near Mount Hood and at Newberry Crater has also found water at unusually high temperatures. This has led to the concept that much of the Cascade Range is underlain by heat sources: magma bodies, which are masked by the much cooler water near the surface.

Figure 9-8 Hot Stuff *– Steam and hot water illustrate the presence of heat from near-surface magma in the Bumpus Hell area at Lassen Peak.*

Table 9-2
Probability of Future Geological Events at Crater Lake

Type/Location	Probability
Renewed volcanic activity within or very near the caldera	Greater than one chance in 330 in 30 years
New volcanic vent near Crater Lake	One chance in 300 to 3,000 in 30 years
Large pyroclastic eruption	Unlikely for many thousands of years
Sudden release of gas	Not considered a significant hazard based on Crater Lake's nature
Catastrophic draining of Crater Lake	Extremely unlikely, but would have disastrous consequences
Large earthquake (M>7) on the WKLFZ*	Unknown, probable recurrence between 3,000 and 10,000 years
Large earthquake (M>8) on the Cascadia subduction zone	Unknown
Volcanic earthquake (M=5)	?

* West Klamath Lake Fault Zone
Source: ***Volcano and Earthquake Hazards in the Crater Lake Region, Oregon***, USGS Open File Report 97-487.

Since the time when pioneers and settlers populated the Pacific Northwest, about half the Cascade volcanoes have been observed in eruption. Such activity indicates that magma bodies have reached the near surface at those locations. Between 1914 and 1921 explosive activity and a lava flow occurred at Lassen Peak's crater. Even more familiar are the recent eruptions of Mount St. Helens in southern Washington. In general, many Cascade volcanoes that have been observed in eruption also tend to display hydrothermal features like those at Lassen. With the discovery of thermal springs on the floor of Crater Lake, it is clear that some source of heat is located below old Mount Mazama's caldera. As we have seen in Chapter 4, the site now occupied by Crater Lake has been the focus of volcanism for many hundreds of thousands of years. Thus, it is quite reasonable to expect more activity in the future – the question is when.

These geological events outlined above could be considered as near-future possibilities in a geological time framework (Table 9-2). Prospects for long-term activity are less sure and depend on many factors, some of which are completely unknown at this time. The entire Cascade Range may be waning if the underlying energy source is the Cascadia subduction zone. If the ocean floor ceases to slide beneath the continental margin, the energy source and origin of magma will also diminish, eventually stopping altogether. Low or nonexistent earthquake activity below the Cascades may indicate a slowing of movement, or no movement at all, along this subduction zone. This would make sense as most of the ocean floor plates have been consumed, indicating subduction is coming to an end. However, the behavior of this subduction zone is not well understood, and such an interpretation may be completely wrong. As an alternative, it may be that movement of the Cascadia subduction zone is sporadic and the present time is simply a quiet period, as some research seems to indicate. Although arguments can be made for either case, the final decision waits on time – geological time.

Note: Much of the material in this chapter is based on the USGS Open File Report 97-487, Volcano and Earthquake Hazards in the Crater Lake Region, Oregon, *1997, by C. R. Bacon, et al.*

Additional Reading

Harris, S. L., 1988, Fire Mountains of the West, Mountain Press, ISBN 0-87842-220-X, 379 pages. A good review of all the Cascade volcanoes including a short discussion of future events in the Crater Lake story.

See additional references by Bacon in the Appendix.

Plate 1 *– This plaque on a glaciated andesite lava flow at* ***Discovery Point*** *is where John Hillman first saw Crater Lake on June 12, 1853. This location was established by William Steel much after the fact.*

Plate 2 *– William Steel started construction of the* ***Crater Lake Lodge*** *in 1909, but it was never finished. In 1989 the lodge was closed due to safety concerns, then completely rebuilt and reopened in 1995.*

***Plate 3 - Chaski Slide**, along the southeastern section of the lake, is the largest example of mass wasting along the caldera walls.*

***Plate 4** – This view of **The Watchman** (L) and **Hillman Peak** across the lake from Rim Village illustrates how Mount Mazama's collapse exposed the interior of these smaller volcanoes.*

***Plate 5 – Mount Scott** represents both young and old geological activity at Crater Lake. The volcanic rock is some of the oldest in the Park, while the "scooped out" cirque here was eroded by glacial activity much more recently.*

***Plate 6 – Phantom Ship**, as viewed from lake level, is part of one of the earliest volcanic features of Phantom Cone.*

Plate 7 – Garfield Peak features one of the most popular hiking trails in the park. Excellent views of the lake, the interior of the caldera, and the Klamath Basin reward visitors who reach its summit.

Plate 8 – This ***prysmatic jointed block*** *of granite, not part of Mount Mazama's rocks, was ejected during the climactic eruption.*

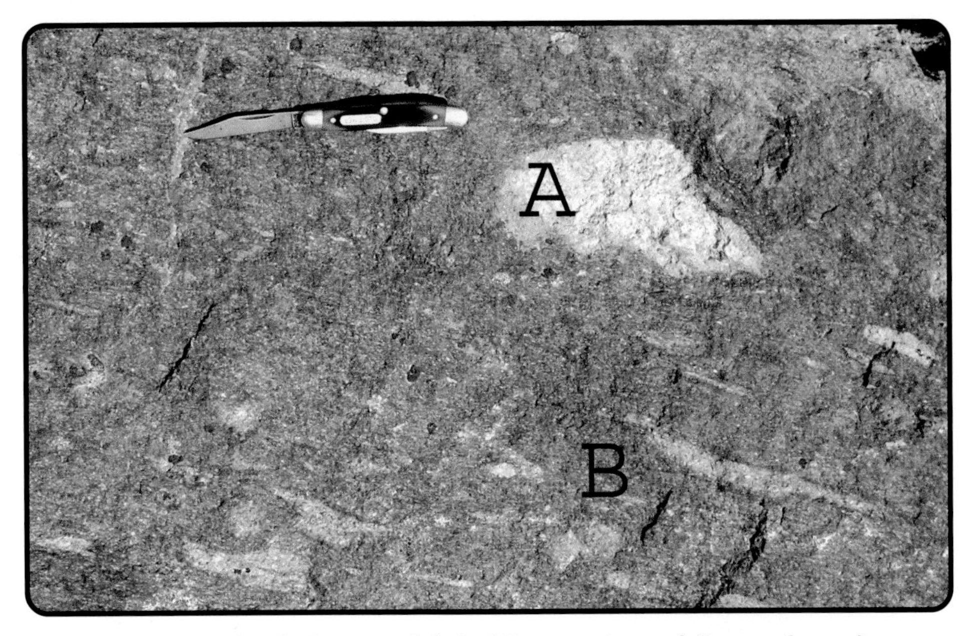

***Plate 9** – The dark stripe labeled B is a piece of **flattened pumice** that was squashed when the ash flow became hot. Its original size would have been about the same as the pumice clast labeled A.*

***Plate 10** – **Wizard Island** represents only the top portion of a much larger feature on the floor of Crater Lake. Devils Backbone and Llao Rock can be seen on the far caldera wall from this view near Discovery Point.*

Plate 11 – The Devils Backbone *This is the largest and most prominent dike along the caldera wall. It stretches from the shoreline nearly to the rim.The lower portion rises about 155m (500 ft) from the water surface and is 15 m (50ft) thick.*

Plate 12 – This bank of ***climactic ash*** *is piled up in the Umpqua River Valley northwest of Crater Lake. The river channels surrounding Mount Mazama were poured full during the climactic eruptions. Later, as the rivers reestablished their valleys, the ash was easily eroded and carried away.*

Plate 13 – *This is a view of the eastern wing of one of the best known features in the caldera wall,* **Llao Rock**. *It clearly illustrates the climactic ash material on top. Air-fall from the explosive event that initiated the eruption, and ultimately produced the massive lava flow that filled its explosion crater, underlays the flow. The Llao Rock lava flow is about 600 m (2,000 ft) thick.*

Plate 14 – The ***Pinnacles*** *in Sand Creek Canyon were formed as fumaroles developed after the climactic ash flows stopped moving. Hot gases escaping up through the ash deposited minerals that "hardened" the material around the vents. Later, as streams reestablished their valleys, the more resistant portions remained standing as pinnacles.*

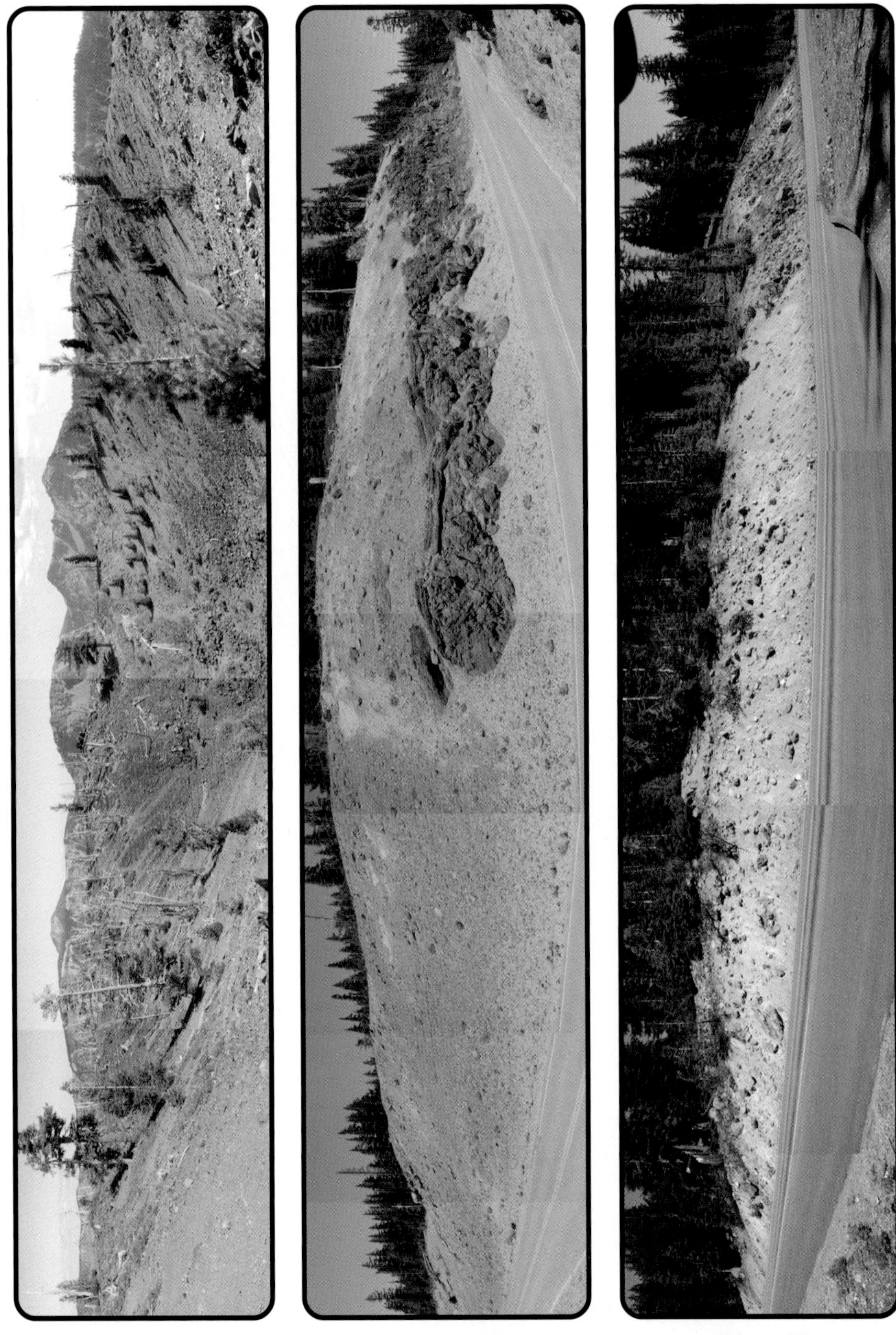

Plate 15 ***Plate 16*** ***Plate 17***

Plate 15 – Wizard Island Crater *The summit of Wizard Island's cone is about 125 m (400 ft) across and 30 m (90 ft) deep. It is typical of many cinder cones, although others do not exhibit such a distinct depression.While Wizard Island appears to have the "ideal" shape of these small volcanoes, the portion visible is only part of the material that forms the cone.It rests on a larger platform beneath Crater Lake's surface (Figures 7-4, 7-5 and 7-6). As discussed in Chapter 7, this feature is young, and part of the final active volcanism in the park. William Steel named many of the features in Crater Lake Natinal Park including Wizard Island, which he thought looked like a wizard's hat.*

Plate 16 – Cleetwood Lava Flow *This roadcut along Rim Drive, just east of the Cleetwood parking area, records the events as the climactic eruption began. The red lava exposed along the right is the top of the Cleetwood flow and was altered (oxidized) by fumarolic gases produced by the eruption. Pumice produced during the climactic eruption accumulated on the Cleetwood lava's surface and was also altered and partially welded (Figure 5-11). This indicates that the climactic event occurred soon after the Cleetwood flow formed. The loose debris at the surface here are proximal ash flow and air-fall deposits produced during the climactic eruption.*

Plate 17 – Corkscrew Roadcut *This exposure between Park Headquarters and Rim Village provides evidence for the collapse of a dome(s) that developed on or near the south summit area of Mount Mazama. See Chapter 5 for a discussion of this exposure (Figure 5-4 and Table 4-1).*

Appendix

Glossary
References
Index
About the Author

Appendix

Glossary

A

aa - lava with a surface composed of angular, jagged blocks.

absolute age - the geological age of a rock, feature, or event using units of time, usually years.

agglomerate - an accumulation of angular fragments of volcanic rocks.

air-fall (ash-fall) - pyroclastic materials, usually fine ash or pumice, injected into the air, falling from an eruption column or ash cloud and later accumulates on the surface.

algae - a group of one-celled or colonial organisms that contain chlorophyl, usually in aquatic environments.

andesite - a dark gray volcanic rock containing 53-63 percent silica of moderate viscosity when in a molten state. Its composition and eruptive character is between basalt and dacite.

ash, volcanic - loose pyroclastic material consisting of fragments (lava or rock) under 2 mm (0.08 in) in diameter, blown into the air by volcanic explosions.

ash flow - an avalanche consisting of volcanic ash and gases, usually at high temperatures, moving down the slopes of a volcano.

auto-brecciation - fragmentation of portions of the already consolidated crust of a lava flow that becomes incorporated in the still-fluid portion.

avalanche (volcanic) - a large mass of volcanic rock materials falling or sliding rapidly downslope under the force of gravity.

B

bacteria (mat) - a group of microscopic one-celled organisms, some forms building up into masses (mats) where hydrothermal waters vent on the floor of Crater Lake.

ballistics/ballistic blocks - rock fragments explosively ejected from a volcano with a ballistic arc.

basalt - a dark colored, low silica magma or lava (less than 53 % SiO_2) erupted at a high temperature. It is relatively fluid (least viscous), thus eruptions are generally nonexplosive and fine grained when cooled.

basaltic andesite - volcanic rock in the silica range between basalt and andesite (~ 55% SiO_2).

bedforms - the dunes-like pattern of pyroclastic flow materials produced during a volcanic eruption.

bench(s) - a narrow, relatively level area bounded by steeper slopes, usually caused by erosion along a beach.

blocks (volcanic) - angular fragments of volcanic material, usually large pieces of a lava flow.

block lava - lava flows displaying a tumultuous mass of angular blocks.

bomb - volcanic fragments of lava which were liquid or plastic when ejected that acquire distinctive shapes of surface features while moving through the air.

breadcrust bomb - a volcanic bomb with crack-like surface features similar to bread crust.

breccia (volcanic) - course, angular, broken volcanic fragments cemented together by finer-grained matrix.

C

caldera - a large basin-shaped volcanic depression, usually circular or oval in shape, formed by the collapse of a volcano due to rapid removal of voluminous magma from below the mountain. Calderas have a diameter many times greater than that of the original crater and are usually a kilometer (mile) or more across.

carbonize - the replacement of (usually) organic materials by carbon.

chronology - arranging events in their proper sequence in time.

cinders - loose, vesicular volcanic ejecta, either fragments of lava or rocks with diameters in the 4 to 32 mm. (½ to 5 in) range.

cinder (scoria) cone - a small cone-shaped monogenetic volcano composed of basaltic to andesitic cinders that form around a vent, usually at an angle of about 32° with the horizontal.

climactic eruption - (Crater Lake) the catastrophic highly explosive eruption of a volcanic cone (Mount Mazama) or feature ending in the collapse of a portion of the volcano to form a caldera.

collapse cloud - a portion of a Plinian cloud that loses energy and falls back down to the surface.

columnar joints - parallel, prismatic columns, often hexagonal or pentagonal in cross section - usually formed in basaltic lava flows.

composite volcano (stratovolcano) - a volcanic edifice formed by the accumulation of many layers of high-viscosity lava and fragmental volcanic debris during repeated eruptions at a central vent or closely spaced vents, commonly forms a high, steep-sided, volcanic cone.

continental drift - the concept that continents move a substratum of the ocean crust.

continental plate - a portion (block) of the Earth's crust underlaying a continental mass that moves as a unit.

convection - the transfer of heat by the circulation or movement of heated portion of a liquid or gas.

convective (gas) thrust - upward movement of volcanic materials caused by the release of hot gases during a Plinian eruption.

cored bomb - volcanic bomb with a nucleus of older lava that has been coated with new lava.

crater, volcanic - a circular, steep-walled depression in the summit or flank of a volcano where lava or pyroclastic material is ejected from a vent.

cross-cutting relationship - the existence of one geological feature that interrupts another demonstrating which one was emplaced first and thus is the oldest.

crust - a term referring to; 1) the outermost layer or shell of the Earth, 2) the solid layer on the surface of a lava flow.

crustal plates - the series of large slowly moving plates that form the outermost layer of the Earth.

crystal fractionation - heavier crystals settle in magma due to gravity forming separate regions of minerals.

crystal-rich - erupted pyroclastic volcanic material with an abundance of crystals.

D

dacite - a light-colored, silica-rich volcanic rock or magma (63-68 percent SiO_2) of high viscosity when in a molten state; eruptions are commonly explosive and may produce voluminous tephra, pyroclastic flows, and lava domes.

debris avalanche - rapid, usually sudden, sliding/flowage of incoherent, unsorted loose fragments of rock material moving downslope under the force of gravity.

debris flow (mudflow) - mass movement, rapid flowage of rock material and water downslope, under the force of gravity.

dendrochronology - study of annual growth of tree rings to determine the dates and chronology of past events.

diapiric rise - the process where magma is thought to rise toward the surface as slender plume-shaped bodies.

diatomaceous earth - fine siliceous earth composed mostly of the cell wall of diatoms.

diatoms - silica shells, either modern or fossil, composed mostly of certain one-celled marine algae.

differentiation - the process of producing more than one rock type from a single magma as it cools.

dike - a tabular rock body, usually igneous, that cut across existing structures or younger existing rock.

dome - (volcanic) a mass of solid rock that forms when viscous magma is erupted slowly from a vent and piles up lava around and over it. Its sides are steep and typically mantled with unstable rock debris formed during and shortly after the dome is emplaced.

E

earthquake - sudden series of vibrations in the Earth's crust caused by the abrupt rupture (fault) of rocks being stressed.

earthquake magnitude - measure of an earthquake's strength (the amount of energy released) determined by seismographic observations.

ejecta (volcanic) - material thrown out by a volcano, pyroclastics.

evaporation - to change from a liquid or solid state into a vapor (gas).

explosion crater (volcanic) - a saucer-shaped to conical depression produced during an eruption.

exsolution - transition of a volatile component (a gas) from solution in magma to a gas phase in the fluid.

F

facies - a portion of single rock unit, volcanic ejecta produced during the climactic eruption, as used here.

fault - fracture in rocks where there has been displacement of opposite sides.

fissure flow - a lava flow from an elongated vent rather than a central vent as in a volcano.

fracture - a general term for any break or failure (crack, joint or fault) in a rock caused by stress.

fumaroles - a vent in the ground associated with volcanic activity where hot water vapor (steam) and/or volcanic gases are emitted to form fumarole pipes and vents.

G

gas thrust phase - the period during a Plinian eruption where ejected materials are propelled upward as if shot from a gun.

gases, volcanic - a group of elements or compounds in the gas state of matter associated with volcanic materials and released during an eruption.

glaciers/glacial - a large mass of ice formed by the compaction and recrystallization of snow that slowly creeps downslope under the force of gravity.

glowing cloud (nuee ardente) - a swiftly flowing and turbulent gaseous cloud, sometimes incandescent and containing ash and other pyroclastic material in the lower part, erupted from a volcano.

GPS - Global Positioning System

graben - an elongate, relatively lowered crustal unit or block bounded by faults along its long margins.

granite - an igneous rock of course grained minerals, primarily quartz and feldspar, that forms (crystallized) below the surface by cooling magma.

groundwater - water accumulated in the open spaces (zone of saturation) below the Earth's surface.

H

half-life - the time required for half of a given specimen of radioactive substance (isotope) to lose half of its radioactivity during decay.

horst - an elongate, relatively uplifted

crustal unit or block bounded by faults along its long margins.

hot springs - a thermal spring with a temperature above that of the human body.

hydrological - the science that deals with continental water (liquid and solid), its properties, circulation, and distribution both on and beneath the Earth's surface.

hydromagmatic - a process where water is heated by magma or mixing of magma and water, often resulting in an explosive eruption due to generated steam.

hydrothermal - pertains to water heated by magma or in association with magma.

I

ice age - a term that refers to extensive glacial activity, usually the Ice Age known as the Pleistocene Epoch.

igneous - a term used for solidified magma, including volcanic rock, also used to describe the processes related to the formation of igneous rocks.

igneous rock - rocks are formed from magma (melted rock) that has cooled and solidified, either within the Earth's crust or on it's surface. Magma that solidifies below the crust cools slowly resulting in coarse-grained (large crystal, ie. Granite) rock. Magma that solidifies at or near the surface cools rapidly and results in fine-grained (with small or no crystals, ie. basalt or obsidian) rock. Igneous rock is composed of mostly silicate minerals.

ignimbrites - rock formed by the deposition and consolidation of ash flows, often called ash flow tuff.

invertebrates - animals without a backbone.

J

juvenile - a term referring to new magma (lava), or pyroclastic ejecta resulting from new magma, that reaches the Earth's surface.

K

K-Ar dating - a radiometric form of dating rocks/minerals using potassium (K) and argon (Ar). The radioactive isotope of potassium (^{40}K) decays to stable argon (^{40}Ar) at a specific rate. Measuring the amount of K and ^{40}Ar present in a rock/mineral allows its age to be calculated. In unaltered volcanic rocks this is assumed to be the time since the magma crystallized.

Katmai - an active volcano in SW Alaska that erupted in 1912 to produce the largest ignimbrite eruption of modern times.

Krakatoa - a volcano located between Java and Sumatra that erupted to produce a large caldera in 1883.

L

lag deposits - a residual accumulation of rock fragments, usually coarse, remaining on the surface after finer materials have been removed by various means (wind, water, etc.).

lahar - an Indonesian word for a rapidly flowing mixture of rock debris and water that originates on the slopes of a volcano. Lahars are also referred to as volcanic mudflows or debris flows. They form in a variety of ways, chiefly by the rapid melting of snow and ice by pyroclastic flows, intense rainfall on loose volcanic rock deposits, breakout of a lake dammed by volcanic deposits, and as a consequence of debris avalanches.

landslide (rockslide) - a general term referring to a wide variety of mass-movement processes under the force of gravity (rockslide, rockfall, earthflow, etc.)..

lapilli - pyroclastic materials in the range between 2 mm and 64 mm (3/64 in - 2 ½ in) in diameter.

lava - a general term for; 1) molten volcanic rock materials (magma) extruded onto the surface, 2) solid rock that solidifies from molten volcanic rock material.

lava tube - a hollow space (cave) beneath the surface of a solidified lava flow resulting from the withdrawal of molten lava after a curst forms on the surface.

lithic (rock) fragments - in the volcanic sense, refers to preexisting solid rock materials.

M

maar - a relative shallow, flat-floored explosion crater with walls composed mostly of loose debris.

mafic inclusions - a small section of iron and magnesium-rich material found in some lavas.

magma - molten rock that may contain suspended crystals and/or vapor bubbles, forms within the upper part of the Earth's mantle and crust and produces lava or pyroclastics when erupted onto the surface.

magma chamber - the location where a reservoir of magma occurs beneath the Earth's surface, often associated with a volcano above the surface.

magma mixing - a process where magmas of two different compositions are mixed together to form a magma with a new composition (a poorly understood process).

magmatic (volcanic) gases - volatile matter held in magma (dissolved) and released during an eruption, typically includes water vapor along with carbon, sulfur and hydrogen gases, and others.

magnetic/magnetized - pertaining to a magnetic or magnetism, having the properties of a magnet.

mantle (Earth's) - zone of the Earth between the base of the crust and above the core, divided into an upper and lower part with a transition zone between.

mass wastage - all processes on the Earth's surface where solid material moves downslope due to gravity.

metamorphic - pertaining to the process of metamorphism or resulting products.

mid-ocean ridge - the massive mountain-like feature extending around the Earth formed on the ocean floor by upwelling of magma (spreading center).

mineralogy - the study of minerals: their formation and occurrence, properties and composition, and thier classification.

monogenetic volcano - a volcano (usually small) constructed by a single eruption or series of similar eruptions closely spaced in time, usually with an active life span of a few months or years.

moraine - a mound or ridge of unsorted and unstratified glacial material, usually till.

mudpots - a type of spring containing boiling mud, usually associated with escape of heat in volcanic areas.

multibeam echo sounder - a sophisticated device used to map water depth using sound signals.

O

obsidian - a dark-colored volcanic glass (rock), usually with a composition like rhyolite.

ocean slab - a portion of the ocean crust adjacent to a subduction zone that is sliding below the adjacent plate at an active plate boundary.

oligotrophic - refers to a lake with a deficiency of plant nutrients and an abundance of dissolved oxygen.

P

pahoehoe - a type of lava flow having a glassy, smooth, and billowy or undulating surface; usually associated with basaltic lavas.

paleomagnetism - magnetic polarization acquired by minerals, especially magnetite, in a rock at the time the rock was solidified or deposited.

parasitic cone - refers to a volcanic cone (crater or lava flow) that develops on the side of a larger cone.

parent magma - a magma from which another magma is derived or from which certain igneous rocks are solidified.

percolation - the process of a fluid (usually water) that passes through porous material, used when goundwater flows through rock or soil due to the gravity.

phenocrysts - the larger, more conspicuous crystals in an volcanic (igneous) rock.

phytoplankton - an aggregate of passively floating or drifting microscopic plant organisms occurring in a body of water.

pillow lava - a pattern observed in some extrusive igneous rocks that resemble a series of pillows stacked one upon another, usually formed when lava flows into water.

pinnacles - 1) a tall, slender, tapering or pointed tower of spire-shaped pillar of rock, 2) in volcanic use, it describes the resistant spires representing the remains of fumaroles.

plate tectonics - a model of the Earth's near surface that is composed of a few large blocks which "float" on a viscous underlayer in the mantle and move with respect to one another.

plinian column - a massive mushroom-shaped cloud produced during a large volcanic eruption as pyroclastic material is blown into the atmosphere by escaping magmatic gases.

plinian eruption - an explosive volcanic eruption producing a steady, turbulent stream of fragmented magma and magmatic gas being released at high velocity, a towering eruption column is produced that rises buoyantly into the atmosphere.

porphyritic - the texture of an igneous rock in which large crystals (phenocrysts) are surrounded by smaller crystals or glass.

postcaldera eruption - an eruption that occurs following the formation of a caldera (such as those on the floor of Crater Lake).

precipitation - discharge of water (rain, snow, hail, etc.) from the atmosphere falling on the Earth's surface.

preclimactic eruption - volcanic eruption(s) volcanic materials (rhyodacite) prior to the climactic event (at Crater Lake about 20,000 years before the climactic eruption of Mount Mazama).

pseudopillows - a pattern in some lava flows that resemble pillow lava structure but is less distinct.

pumice - a light-colored, vesicular, glassy rock commonly have the composition of rhyolite.

pumice blocks - volcanic rocks larger than 64 mm (2 ½ in) composed of pumice.

pyroclastics - a general term applied to volcanic material (ash, cinders, etc.) that have been explosively ejected from a volcanic vent.

pyroclastic flow - a dense, hot mixture of volcanic rock fragments and gases, driven by gravity, that flows down a volcano's flank at high speeds.

pyroclastic surge - a turbulent, relatively low-density mixture of gas and rock fragments, driven by gravity, that flow along the ground surface at high speeds.

R

radioactive/radioactivity - a property of some elements that spontaneously emit radiation resulting from changes in the nuclei of their atoms.

radiocarbon - radioactive carbon (^{14}C) with a half-life of about 5730 years, widely used in dating organic materials.

radiometric - a method of determining the age of an object, usually Earth materials, based on measuring the amounts of radioactive elements they contain.

regional volcanism - volcanic activity represented by widespread, generally basaltic to andesitic monogenetic volcanoes such as cinder cones and shield volcanoes, forming background volcanism around the larger centers of activity (composite volcanoes).

relative dating - the placement of a feature, object, or event in the geologic time scale using techniques (ie superposition, cross cutting, etc.) other than absolute age.

resident time - the length of time it would take to fill an empty lake basin.

resurgent dome - a dome-shaped portion of the floor of a caldera that is uplifted by renewed volcanic activity.

rhyodacite - a silica-rich magma or volcanic rock (68 to72% SiO_2) between dacite and rhyolite, see *silicic magma*.

rhyolite - a silica-rich magma or volcanic rocks (> 72% SiO_2), see silicic magma.

Richter Scale - the numerical values, ranging from 1 to 10, used to express an earthquake's magnitude (intensity).

ring fracture - a cylindrically-shaped, steep-sided fault pattern associated with caldera subsidence.

S

scoria - cindery, clinkery, vesicular, crust on the surface of andesite or basaltic lava due to escaping volcanic gas before solidification, often refers to pyroclastic eject.

secchi disc - a circular instrument lowered into water to measure transparency.

seismic/seismograph - pertaining to an earthquake or Earth vibrations and the instrument used to measure these.

scoria bombs - pyroclastic material in the bomb or block size range with a scoria texture.

shield volcano (cone) - a broad, gently sloping mound composed of numerous overlapping and superimposed lava flows that resembles a warrior's shield or inverted shallow bowl, typically basaltic in composition but may be andesite.

silicic magma - magma that contains more than 63% SiO_2, generally the most viscous and gas-rich, includes dacite, rhyodacite and rhyolite.

spring - a place where groundwater flows naturally from rock or soil onto the land surface or into a body of water.

strata - a sheet-like layer of rock, usually used with sedimentary rocks, but any rock type that is deposited in layers.

stratigraphic marker - a rock or material that represents a specific time or geological event that can be traced horizontally for some distance, used to correlate/date one area with another.

subduction zone - an elongated portion of the Earth's surface where one crustal block (lithospheric plate) descends below another as they converge.

superposition - the order in which rocks or geological materials are placed or accumulated in layers, one above another, with the highest bed being the youngest.

surge deposits - pyroclastic material transported by a rapidly moving, ground-hugging cloud produced during a Plinian eruption.

T

talus (slopes) - rock fragments of any size, usually coarse and angular, laying at the base of a cliff or steep slope.

Tambora - a caldera in Indonesia formed in 1815 by the most massive volcanic event in modern times.

tephra - collectively, all fragmental rock material, including magma, ejected during a volcanic explosion or eruption.

terranes - a group of rocks formed at one location and transported to another location by plate tectonics.

tree casts - a vertical opening in a lava flow remaining after a tree has been incinerated.

thermal vents - locations where volcanic heat, water or gas, escapes at the Earth's surface.

topography - configuration of part of the Earth's surface including relief and position of natural features.

trench (ocean) - a narrow elongate depression of the ocean floor oriented parallel to the trend of a continent or island arc.

tuff - a compact pyroclastic deposit of volcanic ash and dust.

turbidity current(s) - a dense current in air, water or other fluid caused by differences in suspended matter, in water these move down slope, spread out, and deposit transported sediment.

U

U-shaped valley - a round-shaped, flat-bottomed valley eroded out by a glacier.

unconformity - a break in the rock record where time is not represented by rock material, a buried erosional surface.

USGS - United States Geological Survey.

V

vapor-phase crystallization - crystallization of minerals directly from elements in a magmatic gas, often associated with fumaroles.

vent - an opening at the Earth's surface through which volcanic materials are ejected.

volcano/volcanic - a vent or pertaining to a location in the Earth's crust where magma is expelled to produce lava, pyroclastics or magmatic gases.

W

wave-cut platform - a horizontal or gently sloping surface produced by wave erosion extending out from a lake's shoreline.

welding - compaction of a hot pyroclastic deposit under its own weight after deposition.

Z

zoned magma chamber - a model suggesting magma located below a volcano is layered, being silica-rich on top gradually changing to a more mafic composition with increasing depth.

zooplankton - an aggregate of passively floating or drifting microscopic animals occurring in a body of water.

Appendix

References

Popular Publications

Alt, D. D. & Hyndman, D. W., 1995, *Northern Exposures – A Geologic Story of the Northwest*, Mountain Press Publishing Co., ISBN 0-87842-232-0, 443 p. A review of the geological history of the states of Oregon, Washington, Idaho, and Nevada, including parts of Utah, Montana, and Wyoming.

Cranson, K R., 1982, *Crater Lake - Gem of the Cascades, The Geological Story of Crater Lake National Park*, 2nd Edition, KRC Press, 111 p. A review of the park's geology as it was understood at the time and intended for visitors and others interested in Crater Lake geology.

Frances, P., 1983, Giant Volcanic Calderas, *Scientific American*, V. 245 N. 6 (June) p. 59-70. Examines how large calderas develop and reviews several that have formed in the past million years.

Frances, P., *Volcanoes*, 1976, Pelican Books, 368 p. A somewhat dated but excellent little book that presents the concepts and principles of volcanoes and volcanology.

Harris, S. L., 1976, *Fire and Ice – The Cascade Volcanoes*, The Mountaineers – Pacific Search Press, ISBN 0-916891-39-2, 320 p. A illustrated and detailed discussion of the Cascade volcanoes and their geology as they were understood in the mid–1970s.

Keddington-Lang, M., Ed., 2002, Special Issue: Crater Lake National Park at 100, *Oregon Historical Quarterly,* Oregon Historical Society, Spring 2002, V. 103, No.1, 123 p. A collection of illustrated historical essays to celebrate Crater Lake National Park's centennial.

Kittleman, L. R., 1979, Tephra, *Scientific American*, V. 241, N. 6 (Dec.), p. 160-170. A popular discussion of the nature of tephra.

McKee, B., 1972, *Cascadia – The Geologic Evolution of the Pacific*, McGraw-Hill Book Co., 394 p. A summary of the geology of the pacific northwest with emphasis on the Cascade Range.

The Mountain with a hole in the top – Reflection on Crater Lake, 2001, Shaw Historical Library, V.15, Oregon Institute of Technology, ISBN 0-9667919-3-2, 120 p. A collection of first-person accounts and historical comments about Crater Lake National Park.

United States Geological Survey (USGS), 1956, *Crater Lake National Park and Vicinity* (topographic map).

Williams, H., 1954, *Crater Lake, The Story of Its Origin*, University of California Press, 98 pages. The first popular version of Crater Lake's geology based on Howel Williams' fieldwork in the park.

Professional Papers

Atwood, W. W., 1935, The Glacial History of an Extinct Volcano, Crater Lake National Park, *Journal of Geology*, V. 43, p. 142–168. This classic paper reports a thorough study of the glacial evidence and glacial history at Crater Lake.

Bacon, C. R., 1983, Eruptive History of Mount Mazama and Crater Lake Caldera, Cascade Range, U.S.A., *Journal of Volcanology and Geothermal Research*, V. 18, p. 57-115. A report detailing U.S.G.S. geologist C. Bacon's field investigations in the early 1980s. This paper outlines most of his findings that confirm H. Williams' work and adds important new information to the geology of Crater Lake.

Bacon, C. R. and Druitt, T. H., 1988, Compositional evolution of the Zoned calcalkaline magma chamber of Mount Mazama, Crater Lake, Oregon, *Contribution to Mineralogy and Petrology*, V. 98, p. 224-

256. A technical discussion of the chemical evolution of Mount Mazama's magma chamber and a model suggesting the nature of the climactic eruption.

Bacon, C.R., Mastin, L. G., Scott, K. M., and Natherson, M, 1997, *Volcano and Earthquake Hazards in the Crater Lake Region, Oregon*, USGS Open-File Report 97-487,32 p. Outlines potential volcanic and earthquake events that may occur in the Crater Lake area in the future.

Bacon, C. R., Gardner, J. V., Mayer, L. A., Buktenica, M. W., Dartnell, P., Ramsey, D. W., and Robinson, J. E., 2002, Morphology, volcanism, and mass wasting in Crater Lake, Oregon: *Geological Society of America Bulletin*, V. 114, p. 675-692. Report on the caldera floor geology interpreted from survey data gathered using the multibeam echo sounder conducted in 2000.

Buktenica, M., and Larson, G. L., 1990, Ecology of kokanee salmon and rainbow trout in Crater Lake, p. 185-195, in *Crater Lake: An Ecosystem Study*, E.T. Drake, G.L. Larson, J. Dymond, and R. Collier (editors), Pacific Division of the American Association for the Advancement of Science, San Francisco, 221 p. Reports on the fish found in Crater Lake.

Chapin, C. E., and Elston,W. E., Editors, 1979, *Ash Flow Tuffs*, Special Paper 180, The Geological Society of America, 211 p. A technical publication featuring papers on various aspects of ash flows.

Diller, J. S., and Patton, H. B., 1902, *The Geology and Petrography of Crater Lake National Park*, United States Geological Survey Professional Paper 3, 167 p. This publication reported the first detailed geological survey of the Crater Lake region.

Druitt, T.H., and Bacon, C.R., 1986, Lithic Breccia and Ignimbrite Erupted during the Collapse of Crater Lake Caldera, Oregon, *Journal of Volcanology and Geothermal Research,* V. 29, p. 1-31. A report describing selected materials and their distribution that were ejected during the climactic eruption.

______________, 1989, Petrology of the zoned calcalkaline magma chamber of Mount Mazama, Crater Lake, Oregon, *Contribution to Mineralogy and Petrology*, V. 101, p. 245-259. Reviews the composition and nature of Mount Mazama's magma chamber, and suggests a model for its history of eruption.

Kalff, J., 2002, *Limnology: Inland Water Ecosystems*, Prentice-Hall Inc., Upper Saddle River, N.J., ISBN 0130337757, 592 p. An introductory textbook for the study of lakes, rivers, and wetlands as ecological systems.

Kamata, H., etal, 1993, Deformation of the Wineglass Welded Tuff and the timing of caldera collapse at Crater Lake, Oregon, *Journal of Volcanology and Geothermal Research*, V. 56, p. 253-266. Discusses the nature of the Wineglass Welded Tuff with special emphasis on its deformation, emplacement temperature, and its relationship to the timing of the caldera collapse and the Cleetwood lava flow.

Kettleman, L.R., 1973, Mineralogy, Correlation, and Grain-Size Distribution of Mazama Tephra and other Post Glacial Pyroclastic Layers, Pacific Northwest, *Geological Society of America Bulletin*, V. 84, p. 2957 – 2980. Reports on the mineralogy, distribution, and some other properties of the Mazama air-fall tephra as a valuable stratigraphic marker.

Nelson, C. H., etal, 1988, The Mazama climactic eruption (~6900 yr B.P.) and resulting convulsive sedimentation on the Crater Lake caldera floor, continent, and ocean basin, *Geological Society of America Special Paper 229*, p. 37-57. Report on research examining the nature of the Crater Lake caldera floor, especially the sedimentation that has occurred.

______________, etal, 1994, The Volcanic, sedimentologic, and paleolimologic history of the Crater Lake caldera floor, Oregon: evidence for small caldera evolution, *Geological Society of America Bulletin*, V. 106, p. 684-704. A report of current research on geology of Crater Lake's floor that discusses volcanic features and dating of these features.

Richey, J. L., 1979, *Origin of Divergent Magmas at Crater Lake, Oregon*, unpublished PhD thesis, University of Oregon, 209p.

Sigurdsson, H., Editor, 2000, *Encyclopedia of Volcanoes*, Academic Press, ISBN 0-12-643140-X, 1417 p. A comprehensive, but highly technical, source book that covers all aspects of volcanoes and volcanism. The ultimate reference on anything to do with volcanoes.

Smith, R., and Braile, L., 1984, Crustal structure and evolution of an explosive silicic volcanic system at Yellowstone National Park, in *Studies In Geophysics; in Explosive Volcanism: Inception, Evolution and Hazards*, p. 96 – 111. Using data from geophysical methods to propose a model that suggests the presence of a bimodal rhyolite/basaltic source of magma below Yellowstone park that includes future volcanic activity.

Smith, W. D., and Swartzlow, C. R., 1936, Mount Mazama: Explosition Versus Collapse, *Bulletin of the Geological Society of America*, V. 47 p.1809-1330. An old paper that argues for and presents evidence favoring a collapse hypothesis for the creation of the Crater Lake caldera.

Sparks, R.J., Siguardsson, H. and Wilson, L., 1977, Magma Mixing: A Mechanism for Triggering Acid Explosive Eruptions, *Nature*, V. 267, p. 315-318. Discusses a possible process that would explain how two types of magma may be intermingled in a volcanic eruption.

Williams, D.L. and Von Herzen, R.P., 1983, On the Terrestrial Heat Flow and Physical Limnology of Crater Lake, Oregon, *Journal of Geophysical Research*, V. 88, N. B2, p. 1094 – 1104. Report on hydrothermal fluids entering Crater Lake via at least two thermal springs on the deep lake floor, and how these affect the temperature, circulation, stratification, and dissolved solids of the water.

Williams, H., 1942, *The Geology of Crater Lake National Park, Oregon,* Carnegie Institute of Washington, Publication 540. Field research detailing the definitive geology story of Crater Lake National Park.

__________, 1961, The Floor of Crater Lake, Oregon, *American Journal of Science*, V. 259, p. 81-83. Uses the depth soundings made by the U.S. Coast and Geodetic Survey to interpret the general geology of Crater Lake's floor along with a comment on composition of dredge samples.

__________, & Goles, G., 1968, Volume of the Mazama Ash-fall and the Origin of Crater Lake, Oregon, in Andesite Conference Guidebook (A.R. McBirney, Ed.), *Oregon Department of Geology & Mineral Industries Bulletin* 62, p. 37-41. Discusses the calculations used to determine the volume of ejecta produced during the climactic eruption and collapse of Mount Mazama to form the Crater Lake caldera.

Appendix

Index

Appendix

Appendix

Appendix

Appendix

About the Author

Rod Cranson served as an interpretive ranger at Crater Lake National Park for seven summer seasons. In addition, he has made many other extended visits. During those years, and even before, Rod studied the geology of this classic area. Volcanoes and volcanology have been a fascination to him since early in his college studies, and he has attended numerous seminars and workshops to enhance his knowledge. In addition to travel in the Pacific Northwest, Rod has visited many volcanic areas in other parts of the United States as well as Canada, Mexico, Japan, and several Caribbean islands.

Rod received both his Bachelor of Science and Master degrees from Michigan State University in preparation for a career in petroleum geology. However, the opportunity to teach Earth Science in high school led to a career in education, including 27 years in the Science Department at Lansing Community College. He has received numerous awards including: The Distinguished Faculty of the Year Award from Lansing Community College, Master Teacher Award from the College of Education at the University of Texas, and The Michigan Science Teachers Distinguisted Service Award.

Rod has been active in Earth Science Education on the local, state and national levels. He founded the Michigan Earth Science Teachers Association and was also instrumental in establishing the National Earth Science Teachers Association. His years of teaching were shared with graduate studies in geology, including research at Crater Lake. Enjoying Rod's interests in geology, mountaineering, photography and writing are his wife, Sharon, daughter, Keri and son, Kurt.

Many of the graphics included in this book were prepared by Paul Simon, a student in Fina Arts at the University of Michigan. Paul's drawings are on pages 6, 7, 10, 26, 30, 31, 32, 33, 38, 39 49, 54, 79, 93, 106.

Cover concept by Kurt Cranson and design by Sue Harring.

Notes

Notes

Notes

Notes

Notes

Notes

Notes

Notes

Notes

KRC Press
226 Iris Avenue
Lansing, Michigan 48917